I0835397

First established in 2004, the DATA browser book series explores new thinking and practice at the intersection of contemporary art, digital culture and politics. The series takes theory or criticism not as a fixed set of tools or practices, but rather as an evolving chain of ideas that recognize the conditions of their own making. The term "browser" is useful here in pointing to the framing device through which data is delivered over information networks and processed by algorithms. Whereas a conventional understanding of browsing suggests surface readings and cursory engagement with the material, the series celebrates the potential of browsing for dynamic rearrangement and interpretation of existing material into new configurations that are open to reinvention.

Series editors:

Geoff Cox
Joasia Krysa

Volumes in the series:

DB 01 ECONOMISING CULTURE
DB 02 ENGINEERING CULTURE
DB 03 CURATING IMMATERIALITY
DB 04 CREATING INSECURITY
DB 05 DISRUPTING BUSINESS
DB 06 EXECUTING PRACTICES
DB 07 FABRICATING PUBLICS

www.data-browser.net

This volume is produced with support from Association for Art History, Coventry University, Kingston University, Liverpool John Moores University, and London South Bank University.

DATA browser 07

FABRICATING PUBLICS

Mieke Bal
Bill Balaskas
David M. Berry
Ferry Biedermann
Ramon Bloomberg
Natalie Bookchin
Forensic Architecture
Charlie Gere
Alexandra Juhasz
Steven Henry Madoff
Nat Muller
Carolina Rito
Emily Rosamond
Christine Ross
Gregory Sholette
Terry Smith
UBERMORGEN
Santiago Zabala

DATA browser 07
FABRICATING PUBLICS:
The Dissemination of Culture
in the Post-truth Era

Edited by Bill Balaskas
and Carolina Rito

Published by
Open Humanities Press 2021

Freely available at
www.data-browser.net/db07.html

ISBN (print): 978-1-78542-105-1
ISBN (PDF): 978-1-78542-104-4

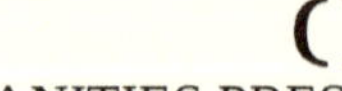
OPEN HUMANITIES PRESS

Open Humanities Press is an international, scholar-led open access publishing collective whose mission is to make leading works of contemporary critical thought freely available worldwide. More at www.openhumanitiespress.org

DATA browser series template designed by Stuart Bertolotti-Bailey.

Book layout and typesetting by Mark Simmonds

The cover image is derived from *Multi* by David Reinfurt, a software app that updates the idea of the multiple from industrial production to the dynamics of the information age. Each cover presents an iteration of a possible 1,728 arrangements, each a face built from minimal typographic furniture, and from the same source code.
www.o-r-g.com/apps/multi

Contents

7 **Acknowledgements**

9 **Fabricating Publics: The Dissemination of Culture in the Post-truth Era**
Bill Balaskas and Carolina Rito

25 **Truth in Transition, as the Decade Breaks**
Terry Smith

43 **Post-Truth and the Curatorial Imperative**
Steven Henry Madoff

49 **Triple-Chaser**
Forensic Architecture

55 **Post-Anthropogenic Truth Machines**
Ramon Bloomberg

73 **Isuma, the Planetary, Listening and the Reinvention of the Counterpublic Sphere**
Christine Ross

96 **Tactical Tutorial for the Post-Internet Era**
Gregory Sholette

99 **Post-Truth as Bullying**
Emily Rosamond

121 **Trust without Evidence: Post-Truth, Culture Policies and Funding Dependency**
Carolina Rito

135 **Artists and Journalists of the World, Unite! A Kitchen Table Polemic**
Ferry Biedermann and Nat Muller

143 **Anarchy near the UK**
Bill Balaskas

153 **Natalie Bookchin in Conversation with Alexandra Juhasz: Performances of Race and White Hegemony on YouTube**
Natalie Bookchin and Alexandra Juhasz

167 **Lying as Truth: Cervantes as Co-Author of Don Quijote**
Mieke Bal

189 **Birth of the Cool**
Charlie Gere

211 **Explanatory Publics: Explainability and Democratic Thought**
David M. Berry

233 **Fabricating Realities (Parkinson Elite)**
UBERMORGEN

241 **Art and the Global Return to Order**
Santiago Zabala

251 **Biographies**

Acknowledgements

We wish to thank Geoff Cox and Joasia Krysa for commissioning this book. Particular thanks are due to Geoff for working closely with us amid the challenging conditions produced by the Covid-19 pandemic. We would also like to thank Mark Simmonds for his design work and Kimberley Chandler for her copy-editing work. Furthermore, we wish to thank Sarah Bennett for her thorough and constructive feedback. This publication would not have been possible without the generous support of London South Bank University and the Association for Art History. Finally, we would like to thank all our contributors for their generosity, patience, and trust.

Fabricating Publics: The Dissemination of Culture in the Post-truth Era

Bill Balaskas and Carolina Rito

The second decade of the twenty-first century has been regularly characterised by the unapologetic merging of subjective personal beliefs and objective facts. The proliferation of "fake news" and "alternative facts" over this period has been credited as one of the most important catalysts for political and social developments around the globe — from Brexit and the election of Donald Trump, to the growth of the anti-vaccination movement. However, the context of post-factuality did not emerge within a void. On the contrary, the manipulation of psychological and social factors for political, economic, or other purposes seems to constitute only part of much larger shifts. Fuelled by the visual language of social media and the economic and political exploitation of data, post-factuality touches on all aspects of our lives by encompassing our sensory landscape: from the graphic interfaces that we use in our everyday online communication, to the algorithmic interfaces that conduct our economic dealings. However, in spite of the omnipresence of such tools and their increasing conflation, their modus operandi is not easily visible — they collectively constitute a "blind spot" in our everyday proceedings.

This publication explores how cultural practitioners, theorists, and institutions might perceive their role within the uncertain landscape of the "post-truth era". The book addresses the multiple challenges posed by the conditions of post-factuality for artists, curators, cultural activists, and their publics: Do cultural institutions have the practical means and the ethical authority to fight against the rise of "alternative facts" in politics, as well as within other aspects of our lives? What narratives of dissent are cultural workers developing, and how do they choose to communicate them? Could new media technologies still be considered as instruments of democratising culture, or have they been irrevocably associated with "empty" populism? Do "counter-publics" exist and, if yes, how are they performed?

In the end, is "truth" a notion that could be reclaimed through contemporary culture?

Fabricating Publics: the dissemination of culture in the post-truth era attempts to answer such questions by including contributions by artists, critics, art historians, media theorists, philosophers, museum curators, and independent cultural practitioners, who explore the multiple — and often contradictory — aspects of post-factuality. Not surprisingly, perhaps, addressing these contradictions begins from the very definition of "post-truth" — a term that is openly questioned or, even, rejected by many of this book's contributors. The inherently problematic prefix "post-", which implies a definitive break in history, or a time when truth remained unquestionable and capable of circulating freely amongst different publics, inevitably raises serious concerns regarding the term's ideological function. In particular, despite the fact that "post-truth" rose to fame in 2016, the conditions that may be used to describe it are older — one might claim, much older.[1] "Fake news" has always existed, and has often defined historical developments in different social, political, and cultural contexts. In addition, it is worth noting that the "fake news" phenomenon and viral misinformation does not merely relate to the dominant narratives and events that Western media, thinkers, and commentators predominantly identify with the rise of post-truth politics.[2] For instance, the right-wing media in Brazil played a key role in the 2016 impeachment of Brazilian President Dilma Rousseff, through the manipulation of both the country's public opinion and the politicians who voted against her.[3]

However, even if we contest its name and span, the so-called post-truth era is likely to leave deep marks on the course of humanity for the foreseeable future — if such a future still exists. Amid several examples, the appointment of climate change denier Amy Coney Barrett to the Supreme Court of the United States by Donald Trump, just a week before the 2020 Presidential Election, is likely to influence US climate policy for many decades to come, at a time when urgent environmental action is needed. A very similar argument could be made in the case of Brazil's President Jair Bolsonaro, whose policies have accelerated the destruction of the Amazon rainforest and its indigenous communities — a global threat to the survival of humanity in the twenty-first century. Most worryingly, perhaps, in the US Election of 2020, and despite Trump's ultimate defeat, more Americans voted for him than in 2016; even though his presidency was defined by a complete disregard for any notion of veracity, Trump received 11,200,000 more votes than he did four years earlier. The fact that Trump's political agenda and decisions were typically based on "alternative facts" seems to have encouraged, rather than discouraged, a very significant proportion of the American electorate — an indication that there have been important shifts in our world since the Great Recession, whose impact we have failed to notice or, at least, to fully apprehend. This publication aspires to contribute to this process of deconstruction and re-appreciation — not least by highlighting the global financial crisis of 2008, and its ensuing fragmentation of the social body, as one of the key catalysts for the rise of the phenomena discussed in its pages.

Most of the contributions included here were written before what has, probably, been the most momentous single event in humanity's recent history — the outbreak of the Covid-19 pandemic. Entering the (currently ongoing) series of national and local lockdowns in the spring of 2020 generated multiple challenges for us as editors, as well as for many of our collaborators, thus delaying the publication of this book by several months. Yet, the arguments made by all of our contributors have a continuous and, in many ways, refreshing relevance, as we begin to see a path out of the pandemic's multiple tragedies. Collectively, the contributions offer a dispassionate and distinctly reflective view, following a period of extreme events on a global scale during which hypermediation has made any discussion around veracity and the role of visuality even more complex. For instance, in her dialogical piece with Ferry Biedermann, Nat Muller argues that asking artists to offer an immediate response to current affairs, as happened during the Arab Spring of 2011 (and on other occasions since then), is an imperative that potentially flattens their art, and does a disservice to the quest for the truth. The situation described by Muller shares many similarities with the cultural responses to the health emergency that we have been experiencing for the last many months, and the myriads of calls to artists and curators to produce work that reflects on the "reality" of the pandemic. As eloquently expressed in the title of a recent article by Cuban-American artist and curator Coco Fusco that focuses on inequalities and authoritarianism, "we need new institutions, not new art".[4]

(Re)building the kind of institutions that Fusco is referring to requires wider action and deeper collaboration — a prerequisite if we are to confront the multiple aspects of post-factuality that strengthen or perpetuate injustices and despotism. The production and dissemination of culture under these conditions has to take into consideration the extraordinary ability of post-truth media to produce a partial view of our world, based on an unprecedentedly complex combination of traditional and new techniques of public manipulation. The immateriality of this media spectacle, at a time of global networked communication, is rendering the formulation of critique and the cultivation of self-reflection into two major challenges for socially engaged cultural practitioners. However, there are also important reasons to be optimistic. In recent years, "contemporary arts institutions and independent curatorial projects increasingly programme around lines of enquiry that go beyond the interpretation and framing of an exhibition's concepts and artworks".[5] Several of the projects, works, and theoretical explorations included in *Fabricating Publics* point towards this expanded role for contemporary cultural praxis. In many ways, such a role may be synopsised through one of the shortlisted words for the Oxford Dictionaries 2016 Word of the Year: "Woke".[6]

Contributions

Terry Smith's essay rejects of the provocations that comprise the statement of intent of this publication. He begins his text by directly addressing one of the key questions that we pose: "Is 'truth' a notion that can be reclaimed through contemporary culture?" Smith responds by refuting the very idea of a "post-truth era", claiming that its use as an umbrella term offers a simplified view of our current condition. Instead, the author puts forward an analysis based on three contemporaneous currents, which form a "meta-picture" of our world: the first consists of efforts to continue, expand, and even totalise modern modes of world-making which began in the sixteenth century; the second is defined by a desire for independence from the dominance of first-current modernities; and the third is the contention between the first two currents, the mixed realities of network cultures, and the impending climate crisis, which have led to the birth of new social movements (Occupy, eco-activism, and anti-globalisation, amongst others). Accordingly, Smith poses a key question of his own in the context of "post-factuality": "What is to be done?" His reply, "Join the Open Strike", is an invitation that is also directed at cultural institutions, that should consolidate their roles as Foucauldian heterotopias, where alternative creativities are produced, preserved, and shared. Starting to think of potential action from our immediate surroundings — as curator, critic and poet Steven Henry Madoff argues — is a suggestion that Smith fully embraces as a path to nurturing a truly revolutionary artistic and curatorial practice.

In his own text, Steven Henry Madoff expands on the need for curatorial and institutional activism, highlighting the fact that, in order to be effective in the "post-truth era", cultural institutions have to ask, before anything else, "whose truth" it is that we are engaging with. Similar to Smith, Madoff turns to Michel Foucault in order to raise the issue of truth's governmentality, and the fact that the scope of our agency is seriously challenged by the technical instruments now deployed to gather and curate knowledge. As an antidote, Madoff asks for "poetic, political, and compassionate retellings that both curate impact narratives and reconstruct our institutions in the name of equity". The author provides several examples of curators and institutions that have been able to resist the normalisation of falsehood: from the resignation of Warren Kanders from the Whitney Museum of American Art's board of trustees thanks to the work of Laura Poitras and Forensic Architecture, to Clémentine Deliss's decolonisation of Frankfurt's Weltkulturen Museum; and from María Belén Sáez de Ibarra's exhibitions in Bogotá about war and the destruction of the Amazon and its people, to Maria Lind's work at the Tensta Konsthall with immigrant communities in Stockholm. Addressing the governmentality of biopower and oppression, alongside the governmentality of truth, is necessary in order to discover our own truths and what they mean for our lives with others.

In their visual essay, Forensic Architecture elaborate on one of the projects that Madoff references as a successful example of how "activism can dynamite the clockworks of power within cultural institutions" in the post-truth context: *Triple-Chaser* (2019). In the work, Forensic Architecture use machine learning, synthetic image generation, and photo-realistic modelling to identify tear gas canisters manufactured by the Safariland Group — a company owned by Warren Kanders, vice-chair of the board of trustees of the Whitney Museum of American Art until the summer of 2019. Given that the sale and export of tear gas from US companies does not appear in public records, it is only through the online circulation of images of tear gas canisters by protesters and activists that you can decipher where such munition has been sold, and who is using it. In order to do this, Forensic Architecture created a digital model of the Safariland-manufactured Triple-Chaser and placed it within thousands of photorealistic "synthetic" environments, "recreating the situations in which tear gas canisters are deployed and documented". Thus, "fake" images helped Forensic Architecture to search for real ones — a distinct "re-appropriation" of the technological tools and methodologies employed by individuals, companies, and governments wishing to manipulate facts. Due to his revealed connections with actions against social movements and civil society, Kanders was forced to resign from the Whitney's board on 25 July 2019. This was an outcome largely catalysed by the exhibition of *Triple-Chaser* at the *Whitney Biennial,* from which Forensic Architecture had withdrawn a few days earlier.

Ramon Bloomberg also explores the ontology of networks and their relation to the fabrication of truth; yet from a different perspective: through tracing the provenance of post-truth "infrastructure" in military operations and war technology. Bloomberg begins his analysis with the largest single loss of life that the Central Intelligence Agency (CIA) had experienced since 1983: the Camp Chapman attack in Afghanistan on 30 December 2009. The author explains how this event shifted the focus of American intelligence institutions "from the anthropogenic traditions of Human Intelligence to new epistemological practices, in which the source of truth is increasingly distanced from the individual human being". Prompted by Arendtian modernity, Bloomberg goes on to connect the aforementioned change in the agency and authorship of truth with a new, oxymoronic temporal order, in which the future precedes the present. This entails that the production and evaluation of the present within large bureaucracies — from the US intelligence apparatus and NASA, to high-end hedge funds and insurance companies — is based on an anticipatory futurity. Within such a speculative environment, truth becomes decontextualised and agnostic. Thus, the pursuit of truth looks increasingly like finding a needle in a haystack. Bloomberg concludes that, "It's not that truth has disappeared from the world, but that access to the production of truth has been displaced". From this point of view, truth-making has, in fact, been a technical arms race.

In her investigation of post-factuality, Christine Ross puts forward a concept that originates from armed conflict: "cyber-balkanization". The term describes the fragmentation of online communities into sub-publics with specific interests, whose function is to systematically avoid viewpoints that are antithetical to their beliefs. And while Madoff asks in his essay, "Whose truth?" Ross adopts another question from anthropologist Nicholas De Genova: "Whose crisis?" The response to the question, as well as to the dangers posed by cyberbalkanization, emerges in Ross's essay about the work of the indigenous Canadian artist collective Isuma. More specifically, Ross analyses Isuma's video and webcast interventions in the Canada Pavilion during the 58th edition of the *Venice Biennale* in 2019. Isuma's work connected climate change with colonial dispossession and migration, through an artistic methodology that has aimed at countering what Dylan Robinson calls "hungry listening" — an extractivist practice, in which settlers absorb what is "digestible" to them in Indigenous culture. Contrary to this, and to cyberbalkanization, Isuma's work is based on the creation of a counter-public sphere "whose modus operandi is to relate worldviews, rather than simply multiply or divide them". Such relationality of perspectives on a planetary scale is, Ross claims, the place where truth emerges. By looking at the distinct perspective of the Inuit community on climate change, as well as at their radicalised dialogical practice of accountability, Ross proposes listening as "the forgotten practice of our times" — a way to weaken post-factuality, through mutual respect, collaboration, acquisition of knowledge, and care for all living beings. These are the conditions necessary for creating effective counter-publics, from both the North and the South, who recognise the primordial crises that they have in common.

In response to these multiple crises, Gregory Sholette offers a "Tactical Tutorial of the Post-Internet Era" by "The School of Dissident Studies". Sholette's sketches illustrate a call for the use of facial makeup to defy facial recognition software. He also calls for organised museum interventions, and the use of DIY stencils, in order to demand a world of fully encrypted emails, un-hackable mobile phones, and cheap anonymous online access. His contribution is subsequently defined as part of a "samiZine", as explained in a Glorypedia entry, which describes how The School of Dissident Studies "revitalised the all-but lost knowledge of simple techniques for creatively disrupting everyday oppression", between the years 2024 and 2031. Sholette's anticipative anti-institutionalism concludes with a mention of the liberation of "post-public spaces held captive by electronically insulated state and paramilitary xenophobe militias" — a poignant call for our age, originating from a not-that-distant dystopian future.

Institutional violence is also a key consideration for Emily Rosamond, who identifies "post-truth as bullying". This is the title of her essay, which opens with the question: "What happens to institutional critique in a moment of flat-out institutional attack?" Rosamond highlights the fact that we are living in a time of multiple crises, when events like Brexit, the Covid-19 pandemic, or the Windrush scandal may be considered as both symptoms and causes of institutional failure. The author connects the crisis of post-factuality with the proliferation of coercion tactics, noting that via networked media these have become more personalised and infrastructuralised at the same time. Established definitions — including legal definitions — of bullying often fail to capture such complexities, thus allowing the figure of the bully to hide behind "normal" institutional practices. This is particularly relevant in workplaces, including art institutions, whose cases of coercion also relate to external pressures such as operating within unfavourable economic and policy contexts. In this post-truth moment of divide and conquer, Rosamond detects the rise of the bully as an anti-charismatic authority, providing the examples of strategist bullies such as Dominic Cummings and Steve Bannon, who stand behind charismatic authoritarians such as Boris Johnson and Donald Trump. Yet, in spite of this "far-right desire to sabotage and dismantle institutions", Rosamond suggests that we should actively rethink institutional practices through the lens of "vice epistemologies"; namely, by studying how intellectual vices take hold within institutions and erode their function. Dismantling the figure of the bully may offer fertile ground for nurturing this collective endeavour.

In her own take on institutional critique, Carolina Rito focuses on the current funding regimes of cultural institutions, exploring how the neoliberal paradigm turned our trust in evidence into a "bureaucratic ruse". Rito begins with a 2014 incident at *The Guardian* headquarters in London, in the aftermath of the Edward Snowden National Security Agency (NSA) leaks, when journalists were asked by the British Intelligence Services to destroy the computers where the leaked files had been stored. Prompted by the irrationality of the demand, at a time when data can be so easily copied and transferred, Rito highlights the fact that the "record" and, therefore, trust lies at the epicentre of neoliberalism's effort to manipulate publics. This pressure on the notion of "evidence" — which has only intensified amid the Covid-19 pandemic — is not unknown to the cultural sector. Ever since the global financial crisis of 2008, cultural institutions have been expected to continue providing the same cultural services, but with significantly fewer resources. Accordingly, museums started diversifying their activities in order to become financially sustainable, through private or other forms of competitive funding. Yet, the neo-positivist approach of "trust in the evidence" that defines "call priorities, or the funding-body strategies, or the ethos of this year's award" radically delimits what cultural institutions are capable of achieving. Similar to the irrational logic of the British Intelligence Services in the case of *The Guardian,* funding bodies create "application forms [that] operate as binding scripts for that future to come; an anachronic prediction". This prescriptive power of funding dependency leads to a form of cultural utilitarianism, within which cultural institutions are treated as mere providers of services for which there is limited provision from other public bodies (e.g. health and well-being). Imbuing the function of the cultural sector with greater freedom to speculate, beyond bureaucratic box-ticking exercises, may help to re-energise the capacity of cultural practitioners to reflect reality against the illusion of the truth-making mechanisms imposed by neoliberalism.

Trust is a concept that also lies at the epicentre of the dialogical text by Ferry Biedermann and Nat Muller. The authors connect the issue with their respective roles as a journalist (Biedermann) and curator (Muller). Biedermann recognises that journalism has been hit hard over the last two decades by falling circulations and viewing figures, which have led to a decline in the mechanisms of scrutiny and accountability. Yet, there is still strong investigative journalism, and this is a fact that artists and curators should not only recognise, but build on. On the contrary, a large part of the art world uncritically adopts the nihilistic view that "journalism equals establishment equals vested interests, cover-ups, bias towards the rich and powerful". Quite ironically, many cultural practitioners ignore, at the same time, that there are multiple conflicts of interest and contradictions in their own community — as Biedermann notes, "yesterday's critic can be today's curator of a major show at the same institution". This often remains hidden behind artivism, which — although inspired by laudable aims — is sometimes based on partial information for the sake of achieving such goals. As a result, artists' claims of "presenting The Truth, singular and absolute" can easily avoid scrutiny. In her response, Muller acknowledges that the often-accusatory stance towards the (mass) media in contemporary artistic and curatorial practices can be counter-productive, as it perpetuates "the idea rehearsed by the likes of Trump, Bolsonaro, and company that the media is not to be trusted and is the enemy of the people". Muller also highlights another aspect of the problem: the art world's obsession with veracity and responding immediately to news stories may damage art's capacity to be reflective and provide more nuanced approaches to social and political issues. In the end, such a hasty interpretation of art's role plays into the hands of neoliberal cultural policies, which are unwisely adopted by many institutions and, subsequently, reflected in the works of artists working within conditions of increasing precarity.

In his visual essay, "Anarchy near the UK", Bill Balaskas offers his artistic perspective on the debate around the role of the press in the context of post-truth politics. Balaskas analyses an installation commissioned by Museu d'Art Contemporani de Barcelona (MACBA) in 2016. In the work, Balaskas created a spatial counter-collage, where all the news stories on the front page of *The Sun* on 25 January 2016 had been cut out and replaced by a series of representative objects displayed in a vitrine. Balaskas left intact only the newspaper's dramatic title and its reference to anarchy, which actually introduces a Brexit-related story: the Calais immigrant "jungle" in France, and the "refugee crisis" in Europe as a threat to the UK. By juxtaposing the newspaper's title with material representations of the removed stories, Balaskas highlighted the absurdity of today's world, in which spectacle has thoroughly replaced facts. As the audience was called to "reconnect" the missing news stories and infer meaning, the work exposed the challenges that relate not only to unveiling the truth, but also communicating it — making it public. In this way, the work also highlighted the responsibility of the viewer as a key agent in this process; an invitation to heightened criticality and active citizenship.

In their "in conversation" text, Natalie Bookchin and Alex Juhasz discuss Bookchin's installation and film *Now he's out in public and everyone can see* (2019). The work expands Bookchin's long-term practice based on YouTube-built video works, presenting vloggers on multiple screens as they recount incidents that involve a famous Black man. Yet, this is a public that is dispersed, with shards of opinions. There does not seem to be a centre; there is no shared or agreed on truth as various narratives merge. In reality, this distinct chorus offers a composite of reactions, responses, reenactments, and descriptions, as well as a racist conspiracy theory, relating to four different individuals: a politician, a golf player, an academic and TV celebrity, and a singer, who are never identified. Notably, many of the vlogs used in the piece were produced shortly after Barack Obama's election — a cause of anxiety to many white vloggers who discuss Black power and success in their videos. Juhasz identifies that Bookchin's YouTube works have been making manual connections through her research and editing process that are increasingly happening through algorithms. Such associations are made by major Internet corporations for profit through fragmentation and manipulation, despite the fact that social media are regularly using a pretext of neutrality and horizontality through their supposed function as "platforms". However, as Bookchin notes, "It's finally become common knowledge that Silicon Valley won't save us". Similar to Balaskas, Bookchin is asking her audience to reflect on the fragments that are presented to them and, through this "editorial" process, on the nature of truth. This is in direct opposition to her protagonists, who "appear less concerned with connecting with others than with broadcasting their own opinions".

Bookchin's multiple "authors" find echoes in Mieke Bal's essay "Lying as Truth: Cervantes as Co-Author of Don Quijote", which offers a comprehensive reflection on the notion of (co-) authorship. Bal's contribution is based on her video installation *Don Quijote: Tristes Figuras (Sad Countenances)* (2019), starting with the assertion that the work is not her own "any more than the literary text on which it is based is Cervantes' own". Co-authorship is seen by Bal as a means of confronting post-truth ramifications, in spite of the fact that she forcefully rejects the term itself as one that is being "abused by systematic liars". Bal further justifies this rejection through the notion of the "post-traumatic". More specifically, she relates this to the multiple tragedies that Cervantes suffered, which, as she argues, should be perceived as a continuous presence both in his life and in the novel. This means that the readers of *Don Quijote* are, in fact, constantly confronted with the question of fiction and truth. Therefore, instead of focusing on posteriority ("post"), the author focuses on synchronicity — an approach that is vividly reflected in her installation, which features various episodes from the novel that are enacted in the contemporary world. In the installation and in her essay, Bal places particular focus on episodes of Don Quijote's pointless attempts to help people, which often result in complete failure and ridicule. The artist uses such episodes as a way to turn her visitors into empathic subjects, nurturing a multiplicity of reactions as they move from one episode to the next in a non-linear, "free" manner. This pluralisation of the viewing experience and its content is, for Bal, a pluralisation of authorship and, accordingly, a pluralisation of the public — a shift from activist art to what Bal calls "activating art"; namely, "art that shakes up complacency, and makes people think on the basis of perception and affect, and perhaps changes their political opinions". Liberating people form the multiple unfreedoms of our time requires, above all else, fabricating publics willing and able to connect to others.

Charlie Gere also examines post-factuality by embarking on a re-evaluation of the past, through a multifaceted account of the life and work of David Bowie. Despite his love for Bowie's art, Gere offers a critical analysis that begins by examining the way in which Bowie was — falsely — elevated into a type of secular saint following his death on 10 January 2016. Notably, this "elevation" mainly materialised through social media and the Internet — a medium whose cataclysmic effect Bowie had predicted as early as 1998. This was the year in which he announced the creation of BowieNet — his own Internet service provider (ISP) — which was accompanied by the solid realisation that networked media had the potential to do unimaginable good as well as unimaginable harm, as he argued during a BBC interview in 1999. In that sense, phenomena like "Trump, Brexit, Gamergate, 'involuntary celibate' ('incel') massacres, ISIS, the scandals of Facebook, Cambridge Analytica, Russian election hacking, 'post-truth', etc.". would have come as no surprise to Bowie. However, identifying something as being seemingly "neutral" — in this case, the Internet's equal potential for good and bad — does not mean that your own role within this new reality is neutral, too. Turning to Hito Steyerl, Gere highlights how Bowie's cultural emergence as a brand new type of icon-hero since the 1970s readily served neoliberalism and right-wing politics on multiple levels. For instance, you could foreground his fascination with Fascism, both in terms of his normalisation of the regime and his equation of Adolf Hitler with television and rock stars. As Gere notes, such proclamations bear remarkable similarities to the rhetoric of Donald Trump — an affinity that is further accentuated by their common understanding of "something profound about our contemporary culture: that everything is image". Gere suggests that Bowie's aestheticism — his self-transformation into a product-surface — echoes the Futurists' "art for art's sake" doctrine. Yet, if we were to turn to the beginning of the previous century in order to seek post-truth's complex roots, we could not fail to notice the path forward offered by Walter Benjamin — a contemporary of the Futurists: the politicisation of aesthetics to fight against the aestheticisation of politics.

In his essay, David M. Berry contributes to the investigation of post-truth's origins from a more technical perspective; namely, by focusing on the "black boxes" of computational systems. Berry juxtaposes the obscure mechanisms of computational capital with what he calls "explanatory publics" — publics that exert their social right to explanation by gaining the necessary knowledge (social, political, technical, economic, or cultural) to hold institutions and their use of digital technologies accountable. Drawing from the modus operandi of key technology corporations such as Google, Facebook, and Apple (a company also briefly examined in Gere's essay), Berry highlights the fact that a new infrastructure of production has been established, within which algorithmic "wrappers" generate an unceasing stream of abstract labour provided by Internet users. As "data is the new oil", companies build their systems in a deliberately user-hostile way, in order to keep the default options of data collection intact. The cynical reason behind this operation is termed by Berry as "neo-computationalism" or "right computationalism" — a corporate approach in which the epistemology of computation is extoled through the fetishisation of the surface. The latter is employed as a distraction from the Weberian "iron cages", in which algorithms trap citizens in order to monitor them, stripping them of their autonomy and capacity to think independently. In this way, software engineering becomes the basis for social engineering. Digital corporations capitalise on this potentiality, selling algorithms that either support or diminish the reasoning capacities of online publics. Thus, the computational leads to "the liquidation of information modalities in 'fake news', conspiracy theories, social media virality, and a rising distrust towards science and expertise, and the rise of relativism". The only way out of this impasse is, according to Berry, the construction of a "left computationalism" by and for explanatory publics, within which "the contradictions of computational capitalism might be laid manifest, and, more importantly, democratically challenged and potentially changed".

A distinctly poetic, ontological reading of Big Data and the nature of truth within networked communication is provided in the visual essay "Fabricating Realities (Parkinson Elite)" by UBERMORGEN. Through a combination of designs and short texts, UBERMORGEN's contribution formulates "a proposal for neurodiverse species within and around otherness". UBERMORGEN highlight the complex ways in which data, nodes, and networks are now inseparable from our everyday lives; yet "mistakes create narratives of dissent and reveal true vulnerabilities". Otherness may arise from these mistakes, which can reveal the true nature of the world surrounding us. However, realising this multiplicity may be more challenging than initially expected, as it requires facing — and, perhaps, transforming — the narcissistic traumata of our networked existence.

The book closes with an essay by Santiago Zabala, who offers a vigorous defence of the role of artists within the condition of post-factuality. Zabala bases his contribution on the assertion that despite the work of systems that seek to frame and tame expression, artists are capable of finding greater freedom through their work than scientists or philosophers. Building on Slavoj Žižek's argument that "postmodern relativism" is not the cause of alternative facts, Zabala claims that rational universalism as experienced in the twentieth century has resulted in totalitarianism, colonialism, and genocide. Post-truth politics perpetuates this order, given that "alternative facts" and "fake news" formulate a rhetoric of ongoing control for right-wing and capitalist powers. Yet, as the author notes, "Science and systematic thought seek to 'rescue us *from* emergencies' improving and preserving our order, but art at its best attempts to 'rescue us *into* emergencies,' creating event and shock". It is through such "events" and "shocks" that we may confront the truth behind emergencies such as climate change, unemployment, or surveillance. Zabala uses works by Pekka Niittyvirta and Timo Aho, Josh Kline, and Dries Depoorter as examples of how contemporary artists participate in global matters, and how art may "rescue the public into the greatest emergency — the imposed absence of emergency that is the result of an authoritarian return to order and realism".

Notes

1. "Post-truth" was pronounced "Word of the Year" by Oxford Languages on 16 November 2016, two weeks after the election of Donald J. Trump as the 45th President of the United States. "Oxford Word of the Year 2016", *Oxford University Press,* accessed 22 January 2021, https://languages.oup.com/word-of-the-year/2016.

2. As explained by Oxford Languages with the selection of "post-truth" as their Word of the Year 2016, the two events associated with their choice were the EU referendum in the United Kingdom and the US Presidential Election.

3. Teun A. van Dijk, "How Globo media manipulated the impeachment of Brazilian President Dilma Rousseff", *Discourse & Communication 11,* Issue 2 (2017): 199–229, https://doi.org/10.1177/1750481317691838.

4. Coco Fusco, "We Need New Institutions, Not New Art", *Hyperallergic,* 26 October 2020, https://hyperallergic.com/596864/ford-foundation-creative-futures-coco-fusco/? utm_content=buffer1005b&utm_medium=social&utm_source=facebook.com&utm_campaign=buffer&fbclid=.

5. Carolina Rito and Bill Balaskas, "Introduction", in *Institution as Praxis – New Curatorial Directions for Collaborative Research,* eds. Carolina Rito and Bill Balaskas (Berlin: Sternberg Press, 2020), 8–31: 10.

6. "Woke" is an adjective signifying staying alert to injustice in society, especially racism. It originates from African-American usage. "Word of the Year 2016: Shortlist", *Oxford University Press,* accessed 22 January 2021, https://languages.oup.com/word-of-the-year/2016-shortlist.

References

van Dijk, Teun A. "How Globo media manipulated the impeachment of Brazilian President Dilma Rousseff". *Discourse & Communication 11,* Issue 2 (2017): 199–229. https://doi.org/10.1177/1750481317691838.

Fusco, Coco. "We Need New Institutions, Not New Art". *Hyperallergic,* 26 October 2020. https://hyperallergic.com/596864/ford-foundation-creative-futures-coco-fusco/?utm_content=buffer1005b&utm_medium=social&utm_source=facebook.com&utm_campaign=buffer&fbclid=IwAR1JoNyFuGv6aDQN2ooQDTR8Gbmy7ho6J0Eyxlxh29pM9uVsFfDqBPbXAzg.

Oxford University Press. "Oxford Word of the Year 2016". Accessed 22 January 2021. https://languages.oup.com/word-of-the-year/2016.

Oxford University Press. "Word of the Year 2016: Shortlist". Accessed 22 January 2021. https://languages.oup.com/word-of-the-year/2016-shortlist.

Rito, Carolina and Balaskas, Bill, eds. *Institution as Praxis – New Curatorial DirectionsforCollaborativeResearch.* Berlin: Sternberg Press, 2020.

Truth in Transition, as the Decade Breaks

Terry Smith

"In the end, is 'truth' a notion that could be reclaimed through contemporary culture?" I take this question, asked by the editors, as my starting point for considering some of the other issues that they raise.

I begin with a proposition, an inference, and a question. The experience of truth is coming to know the irreducibly given state of the world as it really is; that which remains in place even as the most compelling interpretations fall short of describing, let alone accounting for, it. If so, truth is what will be there as our only reliable resource when the current cloud of "truthiness", "alternative facts", "fake news", disinformation, mystification, doubt-sewing, obfuscation and outright lying finally calcifies, and then crumbles. Given that we can posit such a place and time (the "end" in the editors' question), or at least hold out hope for its being arrived at by those who survive the incipient (yet, for many, already current) catastrophe, what, then, does the *transition* to such a state look like now?

We will see that "contemporary culture" is an insufficient tool of reclamation, and that reclamation is an insufficient goal. Yet, we will also see that a truthful grounding will be necessary, if not sufficient, to bring about the coeval communality that itself must ground the world after it has passed through this "post-truth era".

The darkest night

Whenever I think of a transition of this kind and on this scale, I am haunted by the image conjured by Franz Kafka in one of the parables he wrote in 1920.[1] He imagines a figure struggling to hold his ground, against one in front of him pushing back, while behind him another one pushes forward. This is the parable, in its entirety:

> *He has two antagonists: the first presses him from behind, from the origin. The second blocks the road ahead. He gives battle to both. To be sure, the first supports him in his fight with the second, for he wants to push him forward, and in*

> *the same way that second supports him in his fight with the first, since he drives him back. But it is only theoretically so. For it is not only the two antagonists who are there, but he himself as well, and who really knows his intentions? His dream, though, is that some time in an unguarded moment — and this would require a night darker than any night has ever been — he will jump out of the fighting line and be promoted, on account of his experience in fighting, to the position of umpire over his antagonists in their fight with each other.*

All three temporalities need each other to be themselves; yet, Kafka suggests, each wants to live fully in its own time, while knowing that it cannot. "He" (the present) is universal man whose intentions are uncertain, his motivations not fully known even to himself, and who is capable of changing according to the circumstances: unfreedom is like this. He may have lived entirely in desperation, or, more broadly, in a post-feudal society, or somewhere in the plantation system, or somehow survived the war in Europe. In a situation of endemic instability, he strives for advantage, the only currency. Of course, he has a secret dream, a hope that he will escape this temporal treadmill and be accepted by the external force that normally polices the parameters of the eternal struggle between present, past, and future. Then, he would no longer be a slave to this situation. He would become, if not its master, then its overseer, at least of this portion of it. We can extrapolate: The other combatants (the past and the future) have the same dream, and, perhaps, the same random, unlikely chance at compromised salvation.[2]

What is the "night darker than any night has ever been", the time and place in which elevation to at least partial control over one's destiny and that of others might be possible? In a trivial sense, it is a night so dark that the all-seeing guardian is temporarily unsighted. Or it could be, on the contrary, that on which the Messiah comes. We might be among those who believe that He is the only force capable of "promoting" us out of the temporal tangle. If, that is, He decides to come. For Kafka, we are condemned to such yearnings as we await the always deferred outcomes of the operations of the imperfect, irrational, and unknowable laws made by men. Against both, we hold out the ideal of achieving "the most unbridled individualism", which he evokes in another parable. Were we to achieve it, the

Messiah would appear, of course, too late and needlessly: "The Messiah will come only when he is no longer necessary; he will come only on the day after his arrival; he will come, not on the last day, but the very last".[3] Walter Benjamin may have had Kafka's paradoxes in mind as he wrote his last essay "Theses on the Philosophy of History" (1940), in which he imagines a dialectical materialist approach to history as precisely the ability to arrest its flow, to see its shape, to recognise "the sign of a Messianic cessation of happening, or, put differently, a revolutionary chance in the fight for an oppressed past... A historian who takes this as his point of departure stops telling the sequence of events like the beads of a rosary. Instead, he grasps the constellation which his own era has formed with a definite earlier one. Thus, he establishes a conception of the present as the 'time of the new' which is shot through with chips of Messianic time".[4] For Jacques Derrida also, this coming was less a literal, one-time occurrence, more a potential of time itself, which might manifest in flashes and fragments, the "to come" that is happening, somewhere, in many places, right now.[5]

In contemporary conditions, as these thinkers foresaw, no one kind of force is going to elevate us to some time-space outside of the struggle. Indeed, we are hard pressed nowadays to imagine any future with the kind of presence in the present that it had even for Kafka in 1920, or for Hannah Arendt who wrote a brilliant commentary on the parable in 1961, or for many of us until 1989, or 2001, or 2008, or 2016, or 2019 — name your world-changing year, but notice that, like global warming events, they are occurring more often.

Arendt read the parable against its obvious grain; as evoking an exceptional dynamic between thought and action, which, as she puts it, "sometimes inserts itself into historical time when not only the later historians but the actors and witnesses, the living themselves, become aware of an interval in time which is altogether determined by things that are no longer and by things that are not yet". For her, this is the space of revolutionary possibility, of political innovation, of authentic action, most suited to the post-World War Two moment in which she was writing. But it is also an opening to the most important thing: "In history, these intervals have shown more than once that they may contain the moment of truth."[6]

What if "he" becomes "we"? This is a necessary step for us to take. Today, the in front and behind imagery works less well, as the scale of our contemporaneity is now immense. It is the experience of multitudes, it is driven by seemingly infinite differentiation, and time is everywhere refracted. We, here in the contemporary present, are indeed pressed by past forces — which, however occluded and practico-inert, refuse to be vanquished. Instead, they insist on their presence, and strive to occupy the future. Meanwhile, futures seem present to us mostly as projections that have failed and want to come back, to recalibrate, and try again. As these past and future forces keep fighting to fill in our present, we struggle to find even a temporary place in what should be our natural domain. Displacement inside one's own time is an essential paradox of our contemporaneity. If, for Arendt, who was theorising citizenship within modernity, this state of being was, for those who experienced it, an exception to "the weightless irrelevance of their personal affairs",[7] in contemporary conditions we feel its presence more often than not. Along with the frustrations of weightless irrelevance. What truths are contained in this state of pervasive, seemingly permanent exception? For the rest of this essay, I will consider several statements about truth made during the weeks in which it was written. For a supposedly "post-truth era", they are, unsurprisingly, abundant.

Truth on trial

"There are days in Washington lately when it feels like truth itself is on trial. Monday was one of those days". This is journalist Peter Baker, opening an article on the front page of the *New York Times* on Tuesday 10 December 2019. Entitled "In a Swelling Age of Tribalism, The Trust of a Country Teeters", it listed the impeachment hearings in the House of Representatives as presenting "radically different versions of reality", a report from the FBI Inspector General that "punctured long-standing conspiracy theories even as it provided ammunition for others", and the publication of a trove of documents that "exposed years of government deception about the war in Afghanistan".[8] He does not need to mention President Trump's daily Twitter storm of misrepresentations, insults, and lies, most of which are duly reported by an entranced media. Baker does cite a Washington insider as opining: "truth as a concept gets obliterated because people's investment in certain narratives is so deep that facts simply won't get in the way"; and another: "In an atomized age, that allows individuals to retreat to their own storylines, fantasies and tales in which their side is always good and under attack, and the other always craven and duplicitous".[9] Meanwhile, in London, during a debate between the party leaders in the days before the 12 December general election, the press office of the Conservative Campaign Headquarters re-branded its Twitter account "factcheckUK" and posted tweets supporting the Conservative position. A photograph of a four-year-old being treated while on the floor of a public hospital, illustrating Labor Party accusations that the Tories had underfunded the National Health Service, was published in most major newspapers to great outcry, which was soon blunted by fictitious stories suggesting that it was deliberately staged by the boy's mother. The list could go on, and on, and on... (fill in your local examples).

Finding the truth and reporting it to the people has long been the central purpose of communicative media, thus, their being heralded as the "fourth estate", and ritually regarded as an essential civic function within modern societies, joining government, law, and religion. Authoritarian states, including the hybrid versions recently emergent, know this, and repress journalists who pursue their vocation as above all a matter of "speaking truth to power". In democratic societies since the 1960s, as spectacle slowly became pervasive, entertainment value has become the priority of the media industries. It is essential to their economic viability and serves their ideological motivations. Objective, factual information is eclipsed, the truth is rarely glimpsed, until it is no longer expected. Instead, audiences come to prefer the "pseudo-events" tailored to their already shaped dispositions.[10] By no coincidence, trust in politicians has diminished to the point where democratic majorities in many countries are electing anti-politicians, media celebrities who play act the role of genuine populists, in the hope that these people will expose institutional politics as itself a shadow play, and create enough chaos to engender the birth of some other, *any* other kind of governance, which has to be better than *this.* As the other estates implode, the fourth has become the central site of contestation. Unfortunately, this is happening at the same time that the news itself is becoming what George Orwell called "Newspeak".[11]

The perversion of public spheres by official "Newspeak" is not new. Its current extent, however, is itself news. Orwell's *Nineteen Eighty-Four* (1949) became a number one bestseller in early 2017, triggered by the "alternative facts" irreality being spun by the new Trump administration in the White House. Published in 1949, Orwell had Stalin's regime in mind as his main model, but his dystopia prefigured that created in satellite regimes such as that in East Germany during the postwar period. Novelist Anna Funder, author of *Stasiland: True Stories from Behind the Berlin Wall* (2003), notes that "Current estimates have the number of Stasi agents and informers as 1 for every 6.5 people in the country. Under Hitler, it is estimated that there was one Gestapo agent for every 200 citizens, and in Stalin's USSR one KGB agent for every 5830 people. In the 1990s, the West German media called the GDR 'the most perfect surveillance state of all time.' Now this must be qualified, because of what has come after: the GDR was possibility the most thoroughly surveilled state of the pre-internet age".[12] Several societies would now compete for this title: among them, Great Britain, China, and the United States.[13]

Convolution

Yet, against its own grain, the cloud bank of falsifications also reveals truths that had been obscured: "Yes, yes, Trump is the truth about America, because America has been like this forever. White people haven't seen it before, but we have".[14] This is African American artist David Hammons, giving voice to the widely-held perception — articulated searchingly by Ta-Nehisi Coates among several others — that Donald Trump triumphed, in significant part, because of a backlash against the presidency of Barak Obama on the part of those who can see their white rule of the United States coming to an end.[15] Thus, the blanket defence of Trump's presidency by the Republican Party, despite his wild politics; by the Christian conservatives, despite his craven immorality; and by many workers in the old industries, despite his blatant ruling-class interests. There are, of course, many other factors in play, to some of which I will return. There is a crazy truth to Hammons's extrapolation: "You know, the reason we never see aliens is that everyone in the galaxy knows that this planet is a bad planet. They all know to stay away". His recipe for coping, however, slides into the banal: "I used to have a girlfriend who was a dancer. Dancers are always in pain, and she told me the thing to do was to relax into the pain. That's a good metaphor for the time we are living in".[16]

No, it's not. Especially when it is struck by an artist whose work embodies a relentless rejection of relaxation, a rigorous aesthetic, and an uncompromising ethical stance. But every attitude evoked in the previous paragraph, however partial, distorted, or willful, is a recognition that certain powers are in play that are much greater than those within the purview of the speaker, who believes that these powers determine what happens in the world. They seem to operate mainly at a distance — until, suddenly, they are right here. What kinds of world picture are at stake when "post-truth" seems to constitute an "era", and "relax into the pain" is offered as a pathway through it?

I do not believe that "post-truth" constitutes an era. Naming it as such itself distorts truthfulness by substituting a part for the whole, by promoting yet another "everyone knows" exaggeration. Instead, I see its prominence as an outcome of the fact (the truth, I would say) that three broadly distinct ways of world-picturing are in play today (and have been for some decades), and their interaction will continue to shape the foreseeable future. There is no longer (there never was) a dominant, singular world picture that operates as a total, world-defining regime of truth. Nor is there a myriad of "truths", one for each claimant, or cohort of claimants — that is a lazy fallacy that evacuates the very idea of truth. A Cold War scenario used to claim that there were two regimes of truth, Communist and Free World, or North versus South, West and the Rest, or the East, with dependencies, adjacencies (Third Worldism), and outliers. This, too, lacks conviction as an account of what is going on today.

Concurrence

Against these simplifications I have, for some years now, argued that three currents course through contemporary life and thought, isolating modernity's master narratives, and proliferating divisive differences, while at the same time channelling them into a contemporaneous configuration.[17] We cannot see these currents directly, but we can identify their existence precisely by how they cluster our seemingly inchoate efforts to picture the larger World, by how they tend to organise the concepts and terms we use to create coherent worldviews, and by the concrete effects in the real world. These currents are the shapes, the flows, the configurations, and constellations — in a word, the planes that thought constitutes as it thinks itself today. On each of these planes, we find a clustering of similarities and differentiations into three contemporaneous currents. The currents are formed by the attraction of sufficient similarity between some of their elements, while also being separated by polarities of the power differentials within them. This is a magnetic tension that shapes their historical unfolding as currents. They tend to cluster into three constellations, which are in contention with each other. Taken together, they comprise a meta-picture (a "dialectical image") of the World, in the act of world-picturing as it leaps into visibility.

The first current consists of enormous efforts to continue, expand, and even totalise modern modes of world-making; above all, those that, beginning in the sixteenth century, led to the global dominance of European and United States political, economic, and cultural values. Since 1945, these efforts have changed radically, due to internal dynamics and external opposition. General terms such as "progress", "modernity", and "development" have been replaced by names for clusters of forces that seem to define these changes on a global scale. Among these names: Postwar; Cold War; globalisation; Clash of Civilisations; spectacularity; neo-conservatism; neoliberal economics; post-history; invented heritage; remodernisms; Capitalocene; postcontemporary; reactionary resurgence. I see post-truth politics as the most recent form of this recursive kind of modernisation.

In contrast, and in contestation, a second current took shape during this same period. Its major driver was the desire for independence from the dominance of first-current modernities. Its postcoloniality embraces nationalist ideals, while understanding that they will always be in a state of permanent transition. Terms used to name its essential energies includes these: decolonisation; Indigenisation; anti-Orientalist and postcolonial critique; postmodern parody and pastiche; new realisms; multiple modernities; inverse, hybrid modernisations (China, Asian "tigers"); cosmopolitanism; revived fundamentalisms; insurrectionary anarchisms; decoloniality; and post-Communism. I call this current "transitional transnationality". Its dynamic is such that it might equally be named "transnational transitionality".

The third current is much less geopolitical and world-historical in orientation. It is the world picturing of a generation shaped by the failures of policies based on continuing modernities, the contention between the first two currents, the mixed realities of network cultures, and the impending climate crisis. This generation has registered the contemporaneousness of incommensurable master narratives in the first two currents; it seeks self-fashioning within pervasively surveilled and mediated environments; and is experimenting with several new political formations. These include the movement of movements, anti-globalisation; affiliative connectivity (Occupy); open-strike revolutions; eco-activism; planetary consciousness; and the search for multiple coeval commonality.

Kafka's past, present, future triad echoes in the pushing between these currents. The first is, however, seemingly powerful today, reaching the endgame of its five-century regime, and is, thus, residual when seen in the larger historical sweep. It will not recede, however, without an almighty, perhaps world-destroying, fight. The second was prefigured in the early days of colonisation, as resistance to it, and in the achievement of self-rule by local colonists, until, finally, indigenous and *mestizo* independence was reached. This process was sporadic during the nineteenth century (in South America), exploded during the second half of the twentieth (in Africa and Asia, the Central Europe), and is still occurring in many places today (all of the above, plus the Middle East). Its accumulative energies are impacting every other force and formation, as it becomes the dominant current on a world-wide scale, including in the ex-colonial centres. The third current is emergent, a few decades young.[18]

Whatever the long-term prospects for each current, I see distinct kinds of contention operating between them. Dialectical oppositionality is the rule in relationships between the first and the second, as it was throughout the modern period, but it no longer generates syntheses. Forced alliances and temporary accommodations are common, but, as Hammons reminds us in the case of the inheritance of slavery in the United States, these are fragile. So, too, for contention on a global scale, as the current sparring between China and the US demonstrates. The decolonising nations are no longer rushing to "modernise" on Western models: they are developing their own hybrids. The longest arc of history suggests that what I am calling transnationality — the placing of a nation's sense of itself into modes of incessant transition — has already brought us into a condition where the forces driving the elements of the first current, however vocal and vicious they might be at present, are inclining toward decline, while those clustered in the second current are reaching their maturity, and have stronger prospects for growth (even as the First World continues to weaken them, and extract from them as much as it can). The young activists who are propelling the third current resist being implicated into the core presumptions of both currents. They know that the first has nothing for them, although they do incline towards many of the identarian concerns of the second.

The broad outlines of human self-conception, as it is playing out in societies, economies, and cultures in the world today takes, I am suggesting, these three contemporaneous forms. At the same time, and as the physical setting for these developments, what used to be called the natural world, or first nature, is unfolding according to its long-term logics. Yet, these are being severely threatened by human action, historical and present. The assumption underlying modern industrialisation, nation-building and economic exchange — that the natural world is an infinite resource capable of supplying the raw materials of never-ending modernisation — is cast into doubt. A near total majority of the world's scientists have concluded that the Earth can do no such thing; in fact, that it has already been over-exploited to the point where it is showing itself incapable of sustaining currently existing human life, unless major changes are made in the ways we produce and consume energy. All the "truths" most in dispute today can be traced back to this existential crisis. It is the truth that those promoting the climate of "post-truth" most want to occlude.

When mapped, as I have done, as a snapshot of the present configuration of how we conceive of our contemporaneity in relation to each other and to the world, the reign of incommensurable difference, the lack of coevalness in most of our relationships, is all too obvious. Yet, a desire for coevality — not only between humans, but also between us and the natural world — is emerging in the third cluster. Indeed, I believe that it is driving that cluster, and turning the whole of world-picturing discourse and action its way. This turning is what revolution is coming to mean. Everything registered on the chart, I am suggesting, is tending toward those last three words: "multiple coeval commons". This is what will ground the truth that will count for us when the climate-change deniers who have weaponised "post-truth" finally isolate themselves in their citadels, expecting the rest of us to continue to serve them. But we will not, in acts of revolutionary refusal.

What is to be done? Join the open strike

Action must be specific to the forces at play within the place you are in, in the sector or sectors in which you work, and it must fit your capacity. The editors ask whether cultural institutions have the practical means and the ethical authority to fight back against the "conditions of post-factuality". This concedes too much to the reactionary resurgence. The multitudes are voting with their hearts and minds: they attend museums, especially art galleries, in vast numbers, which they trust as among the very few institutions that are not out to deceive them, however strange the offerings within them may sometimes seem to be. (A banana taped to a wall, anyone?)[19] I was recently asked for comment on what museums of modern and contemporary art should collect these days, but turned the question toward why collect? And, of so, how? The answer to which I kept returning was this: *In order to resist the insane self-destructiveness of our prevailing economic and political systems, museums must continue to preserve the actual artifacts of other kinds of creativity, and to exhibit contemporary alternative creativities, thus acting as one of the few open, public resources for the constructive creativities to come.* Precisely because museums are, as Michel Foucault reminded us, "heterotopias of indefinitely accumulating time" they must remain open for that which is to come — to come from the future, from the past as it reasserts itself, and from what is happening now.[20] At the same time, museums must also put in place flexible modes of critically assessing the art — past and present — that flows though their doors, rooms, and projected spaces. *The fundamental, and paradoxical, principle is that we must collect only that art which will always be provisional, always be full of potential, and always be art to come.*[21]

In *Thinking Contemporary Curating,* published in 2012, I listed several priorities for curatorial practice, exhortations that I kept hearing from curators as they urged themselves towards better, more relevant, necessitous curating: "Exhibit art's work. Renounce reticence. Curate reflexively. Build research capacity. Articulate curatorial thinking. Archive the achievements. Reinvent exhibition formats. Turn the exhibitionary complex. Proliferate alternative exhibitionary venues. Activate infrastructure. Embrace spectatorship. Curate contemporaneity in art and society — past, present, and to come — critically".[22] Six years later, Steven Henry Madoff posed this challenge to curators: "What is to be done in your country, in your political and cultural situation, from your point of view?" After repeating the urgencies of 2012, I added some more.[23] Understand that the accelerationism of shock-doctrine capitalism is a sign of that system's implosion and its replacement by an even more chaotic state of affairs. Support the art and join in the activism against this regime and for coeval communality. Become part of the open strike. Do so in your own city, town, or country while being aware that multitudes are doing the same thing all over the world. An open strike is neither a limited withholding of labour until the bosses come around or the union capitulates, nor is it a general strike of all workers against the state. It is a way of continuing to offer minimal services as usual by some workers, which frees the majority to join in the demonstrations on the street and the many other kinds of work — including artistic and theoretical — necessary to bring the commons into being. Right now, the cultural sector of Lebanon, notably a large group of arts organisations in Beirut, are showing the way. The idea of the open strike goes to the heart of both artistic and curatorial practice when they are undertaken in a truly critical, that is to say, *revolutionary* spirit.[24]

As the decade breaks

I am writing these notes during the last months of 2019, as the second decade of the twenty-first century breaks open, having been in transit between election-eve in not-so-Great Britain, a United States that has impeached its President for only the third time, and I am now in Sydney, which swelters in a smoky haze from the unseasonable, global warming-induced bushfires that surround the city. I am reading Maxine Beneba Clarke's poem "When the Decade Broke" (2020).[25] It captures much of what we need to know about truth today. It also offers a pertinent counterpoint to the Kafka parable.

She begins by evoking the widespread fear, ten years ago, that the world's computer networks would not be able to cope with the dawning of 2000, the false fear of a digital meltdown. For her, however, such a moment would be revolutionary: "in the new century, we, the workers, would be king". She then shifts to now:

just like one day
we'll say

where were you,
   on december thirty-first,
two thousand and nineteen

— and perhaps more importantly —

   who were you
before the decade turned

don't look at me like that,
you know what I mean:
who were you, when thunder was made
   from our protesting children's feet

when 45,
(the then-president
of the united states
of america)
   had just been impeached

we'll say to the young ones

   unthinkable now,
isn't it
    that back then, in this city,
women's bodies were sometimes found
   naked, from the waist down
we would gather in the parks,
for candlelight vigils

         in this very place, the decade
before revolution came,
nobody *led*
though four prime ministers
    rose, and fell;

innocent black folks were shot
    at point blank range
regularly

  across the world
and often incarcerated,
    for no valid reason at all

don't avert your eyes from mine:

you should know
what this place was:

earth on fire,
  from the redwoods of california
to australia's east coast

my god,
    the furnaces
that burned

in brazil, they lost a good part
of the amazon

the sea drew back,
    and tsunamis lashed out
in samoa and sumatra;
sulawesi and nagasaki

in the new decade, we will say
the world
was not always this beautiful way:
in some countries
  small children starved to death
every single day

but all that slowly started to change

and powerful men
were brought to trial
  for heinous acts of hate

we threw them out,
    and relegislated

(they'd made the churches
far more powerful
than the state)

for a good while there
we thought we were doomed

that it was all just way too late

but the decade turned

the decade turned,
 and suddenly,

we were wide awake

lined along the gun-powdered foreshore
faces tilted to the sky

watching the revolution break

Notes

1. Franz Kafka, "Notes from the year 1920", in *The Great Wall of China* (New York: Schocken Books, 1946). Also, in Franz Kafka, *The Great Wall of China and Other Short Works* (Harmondsworth: Penguin Classics, 1991). I used this reference as a gateway to thinking about the three currents operative in contemporary thought and art in Hal Foster, ed., "A Questionnaire on 'The Contemporary': 32 Responses", *October #130* (Fall 2009): my essay 46–54; republished as Chapter 2 of *Art to Come; Histories of Contemporary Art* (Durham, NC: Duke University Press, 2019).

2. There is also a quite specific, local reading of the parable. Scott Spector argues that Kafka, along with others in his small circle of German-speaking Jews born in Prague between 1882 and 1892, "stood apart from the generations before them, as well as the contemporary writers outside of their milieu [...] By pure force of unusual circumstance, they were highly conscious of a discourse they were in the midst of, one that posited the separation but mutual necessity of the figures of author, text, and nation". Scott Spector, *Prague Territories: National Conflict and Cultural innovation in Franz Kafka's Fin-de-Siècle* (Berkeley: University of California Press, 2000), 235–36.

3. Franz Kafka, *Parables and Paradoxes* (New York: Schocken Books, 1961), 81.

4. Water Benjamin, "Theses on the Philosophy of History", in *Walter Benjamin, Illuminations* (London: Fontana, 1973 [1940]), 265.

5. Jacques Derrida, *Specters of Marx: The State of the Debt, the Work of Mourning and the New International* (Chicago: University of Chicago Press, 1994); and the interview following 9/11 in Giovanna

Borradori, ed., *Philosophy in a Time of Terror: Dialogues with Jürgen Habermas and Jacques Derrida* (Chicago: University of Chicago Press, 2003).

6. Hannah Arendt, *Between Past and Future* (New York: Penguin, 2006 [1961]), 9.

7. Arendt, "Between Past and Future", 3.

8. Peter Baker, "In a Swelling Age of Tribalism, The Trust of a Country Teeters", *New York Times,* 10 December 2019, 1.

9. Baker, "In a Swelling Age of Tribalism", 16. See also Mishiko Katutani, "The 2010s were the End of Normal: A Decade of Distrust", *New York Times,* 27 December 2019, https://www.nytimes.com/interactive/2019/12/27/opinion/sunday/2010s-america-trump.html.

10. Daniel J. Boorstin, *The Image: A Guide to Pseudo-Events* (New York: Knopf Doubleday, 2012 [1962]) is cited by Guy Debord, *The Society of the Spectacle* (New York: Zone Books, 1994 [1967]); both inspired Jonathan Crary, *24/7: Late Capitalism and the Ends of Sleep* (London: Verso, 2013). For an acute contemporary analysis see Greg Jackson, "Vicious Cycles: Theses on a philosophy of news", *Harper's Magazine,* Vol. 340, No. 2036 (January 2020): 32–40.

11. George Orwell, *Nineteen Eighty-Four* (Chicago, IL: Houghton Mifflin Harcourt, 2017 [1949], 5.

12. Anna Funder, "Stasiland and Now", *The Monthly* (December 2019–January 2020), 27; and *Stasiland: True Stories from Behind the Berlin Wall* (London: Granta, 2003).

13. Shoshan Zuboff, *The Age of Surveillance Capitalism: The Fight for a Human Future at a New Frontier of Power* (New York: Public Affairs, 2019).

14. Cited by Calvin Tomkins, "The Enchanter", *The New Yorker,* 9 December 2019, 61.

15. Ta-Nehisi Coates, *We Were Eight Years in Power: An American Tragedy* (Harmondsworth: Penguin UK, 2017).

16. Tomkins, "The Enchanter", 61.

17. For example, Terry Smith, *Contemporary Art: World Currents* (London: Laurence King and Pearson Prentice Hall, 2011); *The Contemporary Composition* (Berlin: Sternberg Press, 2016); and *Art to Come: Histories of Contemporary Art* (Durham, NC: Duke University Press, 2019).

18. The dominant-residual-emergent model was first set out by Raymond Williams, "Base and Superstructure in Marxist Cultural Theory", *New Left Review,* No. 82 (November–December, 1973), https://newleftreview.org/issues/I82/articles/raymond-williams-base-and-superstructure-in-marxist-cultural-theory.

19. I am referring here to Maurizio Cattelan's work *Comedian* (2019), which attracted worldwide media attention when it was sold for $120,000 USD at *Art Basel* in Miami Beach, in December 2019.

20. Michel Foucault, "Of Other Spaces (1967), Heterotopias", Architecture/Mouvement/Continuité (October 1984), https://foucault.info/documents/heterotopia/foucault.heteroTopia.en, accessed 8 September 2018.

21. Terry Smith, "Why Collect Now? If So, How? Challenges to Modern and Contemporary Art Museums", in *What Do Museums Collect?,* ed. Sunhee Jang (Seoul: National Museum of Modern and Contemporary Art, 2020).

22. Terry Smith, *Thinking Contemporary Curating* (New York: Independent Curators International, 2012), 256.

23. Terry Smith, "Against Rogues: Curating in States of Crisis", in *What About Activism?* Steven Henry Madoff ed. (Berlin: Sternberg Press, 2019), 71–82.

24. The principles of an open strike are articulated in the call of 25 October 2019, see "Statement on

Open Strike in the Cultural Sector in Lebanon", *Arab Image Foundation,* Issue #2019.04, https://stories.arabimagefoundation.org/issue-201904. See also Hrag Vartanian, "Lebanese Art Community Joins Street Protest", *Hyperallergic,* 25 October 2019, https://hyperallergic.com/524105/lebanese-art-community-joins-unprecedented-protests. For a century-old debate on this question, see Georges Sorel, *Reflections on Violence* (Cambridge: Cambridge University Press, 1999 [1908]); and Walter Benjamin, "Critique of Violence", in *Reflections: Essays, Aphorisms, Autobiographical Writings* (New York: Schocken Books, 1986 [1921]). I extrapolate the concept for curatorial practice in Terry Smith, *Curating the Complex and The Open Strike* (Berlin: Sternberg Press and Cambridge, MA: MIT Press, 2021), first single-essay publication in *Thoughts on Curating* series, edited by Steven Henry Madoff.

25. Maxine Beneba Clarke, "When the Decade Broke", *The Saturday Paper*, 21 December 2019–24 January 2020, 38.

References

Arab Image Foundation. "Statement on Open Strike in the Cultural Sector in Lebanon". Issue #2019.04. https://stories.arabimage foundation.org issue-201904.

Arendt, Hannah. *Between Past and Future.* New York: Penguin, 2006 [1961].

Baker, Peter. "In a Swelling Age of Tribalism, The Trust of a Country Teeters". *New York Times,* 10 December 2019.

Beneba Clarke, Maxine. "When the Decade Broke". *The Saturday Paper,* 21 December 2019–24 January 2020.

Benjamin, Walter. *Illuminations.* London: Fontana, 1973 [1940].

Benjamin, Walter. "Critique of Violence". *In Reflections: Essays, Aphorisms, Autobiographical Writings.* New York: Schocken Books, 1986 [1921].

Boorstin, Daniel J. *The Image: A Guide to Pseudo-Events.* New York: Knopf Doubleday, 2012 [1962].

Borradori, Giovanna, ed., *Philosophy in a Time of Terror: Dialogues with Jürgen Habermas and Jacques Derrida.* Chicago: University of Chicago Press, 2003.

Coates, Ta-Nehisi. *We Were Eight YearsinPower:AnAmericanTragedy.* Harmondsworth: Penguin UK, 2017.

Crary, Jonathan. 24/7: *Late Capitalism and the Ends of Sleep.* London: Verso, 2013.

Debord, Guy. *The Society of the Spectacle.* New York: Zone Books, 1994 [1967].

Derrida, Jacques. *Specters of Marx: The State of the Debt, the Work of Mourning and the New International.* Chicago: University of Chicago Press, 1994.

Foster, Hal, ed. "A Questionnaire on 'The Contemporary': 32 Responses". *October #130* (Fall 2009).

Foucault, Michel. "Of Other Spaces (1967),Heterotopias".*Architecture/Mouvement/Continuité* (October 1984). https://foucault.info/documents/heterotopia/foucault.hetero Topia.e, accessed 8 September 2018.

Funder, Anna. *Stasiland: True Stories from Behind the Berlin Wall.* London: Granta, 2003.

Funder, Anna. "Stasiland and Now". *The Monthly* (December 2019–January 2020), 27.

Jackson, Greg. "Vicious Cycles: Theses on a philosophy of news". *Harper's Magazine,* Vol. 340, No. 2036 (January 2020): 32–40.

Kafka, Franz. "Notes from the year 1920". In *The Great Wall of China.* New York: Schocken Books, 1946.

Kafka, Franz. *Parables and Para-doxes.* New York: Schocken Books, 1961.

Kafka, Franz. *The Great Wall of China and Other Short Works.* Harmondsworth: Penguin Classics, 1991.

Katutani, Mishiko. "The 2010s were the End of Normal: A Decade of Distrust". *New York Times,* 27 December 2019. https://www.nytimes.com/interactive/2019/12/27/opinion/sunday/2010s-america-trump.

Madoff, Steven Henry, ed. *What About Activism?* Berlin: Sternberg Press, 2019.

Orwell, George. *Nineteen Eighty-Four.* Chicago, IL: Houghton Mifflin Harcourt, 2017 [1949].

Smith, Terry. *Thinking Contemporary Curating.* New York: Independent Curators International, 2012.

Smith, Terry. *Contemporary Art: World Currents.* London: Laurence King and Pearson Prentice Hall, 2011.

Smith, Terry. *The Contemporary Composition.* Berlin: Sternberg Press, 2016.

Smith, Terry. *Art to Come: Histories of Contemporary Art.* Durham, NC: Duke University Press, 2019.

Smith, Terry. "Why Collect Now? If So, How? Challenges to Modern and Contemporary Art Museums". In *What Do Museums Collect?* Edited by Sunhee Jang. Seoul: National Museum of Modern and Contemporary Art, 2020.

Smith, Terry. *Curating the Complex and The Open Strike* (Berlin: Sternberg Press and Cambridge, MA: MIT Press, 2021).

Sorel, Georges. *Reflections on Violence.* Cambridge: Cambridge University Press, 1999 [1908].

Spector, Scott. *Prague Territories: National Conflict and Cultural innovation in Franz Kafka's Fin-de-Siècle.* Berkeley: University of California Press, 2000.

Tomkins, Calvin. "The Enchanter". *The New Yorker,* 9 December 2019, 61.

Vartanian, Hrag. "Lebanese Art Community Joins Street Protest". *Hyperallergic,* 25 October 2019. https://hyperallergic.com/524105/lebanese-art-community-joins-unprecedented-protests.

Williams, Raymond. "Base and Superstructure in Marxist Cultural Theory". *New Left Review,* No. 82 (November–December 1973). https://newleftreview.org/issues/I82/articles/raymond-williams-base-and-superstructure-in-marxist-cultural-theory.

Zuboff, Shoshan. *The Age of Surveillance Capitalism: The Fight for a Human Future at a New Frontier of Power.* New York: Public Affairs, 2019.

Post-Truth and the Curatorial Imperative

Steven Henry Madoff

Needless to say, we remain in a startling moment of inflection in the political life of the United States and elsewhere. With Benjamin Moser's recently published biography of Susan Sontag, I was reminded of her book-length essay, *Regarding the Pain of Others* (2003), in which she makes the claim that photography is the most incisive and vivid tool to comprehend the misery and violence people afflict on one another.[1] Yet, our present moment has gone beyond the truth she evidently thought was still inherent in photographs, though, even then, it was already becoming a Photoshopped world. Now, nearly twenty years after her essay, ours is a world so pervasively digitally revised that "post-truth" is assumed to be the chilling given — as if truth were an encrusted geological past layered under sheets of dissemblance that excavations can only display as artifacts of its death. The question of whatever happened to truth as a merely historical consideration is nothing less than unforgivable surrender.

From the perspective of curatorial and institutional activism, reflecting on this siege on truth that may have lessened since Trump's departure but remains a national urgency — with other countries suffering their own bullying, autocratic regimes of receding, pseudo-, and anti-democratic distortions — brings with it the question of what these authoritarian tactical narratives reveal. The violation of truth — in other words, the violence of post-truth — produces what can only be called the daily habit of injury to our collective moral life. Having reached what may seem the apogee of the strategic use of disinformation, the violation of truth puts each of us under a droning burden of pressure that collapses our sense of the world, and, therefore, the ways by which we come to know ourselves, at least in part. What happens to our self-knowledge and our relation to others, breathing the daily air of disinformation? But, of course, this question sits along a causal route, and to speak of the regimen of post-truth comes with other questions, among them: Who is it that determines and produces this regimen? Who is the sovereign of this regime of post-truth — though we all know that the continuance of the sovereign is the eternal intention, overriding the health of each of us as citizen-subjects. And we all know who that sovereign is in our own countries.

Digital means may propagate the fields of distortion in our time, but what can be called the "sovereign narrative", to abbreviate centuries of manipulation, has certainly been a concern since Machiavelli's *The Prince* (1513) — what Michel Foucault speaks of as the issue of governmentality. By governmentality, he includes a sinuous complex of strategic actors: "the ensemble formed by institutions, procedures, analyses and reflections, calculations, and tactics that allow the exercise of this very specific, albeit very complex, power that has the population as its target, political economy as its major form of knowledge, and apparatuses of security as its essential technical instrument".[2] When we interrogate "truth", we must immediately ask *whose* truth, whose exercise of subjectivity is imposed on populations as "objective", as a foundational apparatus of control? This can only be broadened today, with technical instruments now deployed to gather knowledge exceeding the range of political economy, or dilating it to the extent that we have to ask what is the scope of agency left to us to exercise? And as I have already said, how does this regimen, this sovereign onslaught of distortions and plain-faced lies twist us and re-form us? How do we sustain ourselves? What *is* my truth?

But I am actually speaking of truth here as an effect whose interrelated causes include the experiential, the semiotic, and the reflective. What we see, what we hear, what we sense around us and react to constellate an embodied knowing in the world. We cognise this knowledge and add to it the semiotic input of cascading streams of data, news, and information that each of us must contend with, filter, interpret, be vulnerable to, believe in, be tricked by, negotiate. And then, there is the internalised reflection on these many inputs that seem to us both solid and less so: chimerical, manipulative, destructive. How does our sense of self, the way we take account of ourselves, coincide, and interact with this multiplicity, this register of signals and noise? As well, this personal form of accounting is based on our private and environmental histories — of biology, family, of lifelong relations with others, and, of course, of that governmentality of biopower, including oppressions of many kinds, that Foucault describes. The Shakespearean precept "to thine own self be true", based on the Augustinian concept of self, and summarised by Polonius in the existential minefield of *Hamlet,* suddenly seems fabulously unreachable, as antiquated and unwieldy as iron amid the global sluices of silicon and optical fiber.

That is to say, I *think* I know myself. Yet, my historical undergirding of Western belief that the self, *my* self, is sui generis and based on an ancient notion of that brittle ultimacy of an unyielding interiority of being, now faces the rich profligacy of lying means — whoever your sovereign is that wields them, who determines and regulates, who offers at once support and threat, who provides but, more likely, denies that old Kantian dream of cosmopolitanism and hospitality, while that Kantian dream itself is based on presumptions of power that must be questioned and revised. In the era of post-truth, the crowd is easy to turn. A melded media beast, and the body politic understood as an assembly of individuals still seeking distinctness within, and outside of, juridical limits, means that we must urgently address collectivity, civility, and security as an opening out of the question of tolerance, whose etymological roots in Latin and Greek are "to suffer, to bear, to endure", while also related "to lift up".

If we can say that the governmentality of truth — now under assault — could not be clearer, we can also say that it is no longer clear how to lift up, how to overcome the wearying indeterminateness brought on by countless lies;[3] what it is to be me, to be us and we, when the campaign of post-truth fills us too often with immobilising disbelief. But I can at least attempt an answer when speaking of the activist role of curators inside museums and other cultural spaces. Here, we can only respond to the sovereign narrative with what can be called "impact narratives" of exhibition-making and other curatorial projects, which repossess truth and reassert compassion for the pain of others, not ridiculed and degraded by the gluttony of power-for-power's sake.

Two years ago, we saw that activism could dynamite the clockworks of power within cultural institutions, as we witnessed with the resignation of Warren Kanders from the Whitney Museum of American Art's board of trustees under the harsh illumination of Laura Poitras and the London-based group Forensic Architecture's scathing video *Triple-Chaser* (2019). This ten-minute video created an outcry, not merely among artworld denizens, but among a broader audience of civil advocates for the right of ethics to help determine the sources of funding for our institutions. The effect of activism by artists and curators resonated all the more globally in the case of mus-eums rejecting the Sackler family's philanthropy built on their OxyContin fortune, tainted as it is by disastrous greed ruining lives — indeed, the most drastically ironic retelling of "regarding the pain of others" by driving pain into addiction and death. The more recent pressure on Leon Black to step down as chairman of the Museum of Modern Art in New York is yet another example of activist energy.

Yet, we have to go beyond what is called in the jargon of the artworld "institutional critique", though important investigative works by Hans Haacke and caustically parodic performance pieces by Andrea Fraser over the last several decades have pointed the way toward revealing multifarious forms of cynicism and corruption embedded in cultural institutions. Works like Poitras's continue this line. Now, we have to direct an outward trajectory that confronts the blanketing storm not only of culture but of the world's landscape of rule of law and its dereliction, of clarity and integrity among those elected to lead, and the passion for economic equity and social justice. The breathtaking deviousness and violence that manipulate a fundamental vision of security and commonality, that undermine any sense of knowing oneself and blatantly twist exploitation as "for the people", are the subjects of artists and curators as citizens who have to take back the telling of life. This, too, was confronted at the Whitney when its curators displayed Dana Schutz's rendering of Emmett Till's racist murder in her painting *Open Casket* and the controversy it caused at the *2017 Whitney Biennial.* The legacy of Okwui Enwezor — including such acts of contemporary retelling as *The Short Century: Independence and Liberation Movements in Africa: 1949–1994* and the eruption of artists' voices in his 2002 edition of *Documenta11,* enunciating their locally expansive uniqueness and a transnational community of emancipatory visions — offers powerful models for impact narratives.

But there are so many more poetic, political, and compassionate retellings that both curate impact narratives and reconstruct our institutions in the name of equity. I think of Maria Lind's aim to invite and imaginatively engage the underserved immigrant community in her Stockholm neighbourhood through inclusive social programming she did when she directed the Tensta konsthall. Of Clémentine Deliss's revamping of Frankfurt's Weltkulturen Museum, demanding that centuries of colonial cultural pillage of African objects be reconsidered in contemporary unfoldings of the truth by bringing artists, writers, philosophers, and others into the museum's ethnographic holdings to produce exhibitions and discussions that yielded both visual surprise and critical recontextualisations. Of María Belén Sáez de Ibarra's immensely powerful exhibitions in Bogotá, which combine deeply researched documentation, blunt imagery, and historical excursions into mythology and alternative ontologies that have brought raw and nuanced insights to the murderous fifty-year war in Colombia and the ruin of the Amazon and its peoples.[4]

No time in human history has escaped the storms of explosive misery wrought by the autocrats of plunder and repression. But ours is the moment in which the insidious rerouting of truth is now technically reliable on a simultaneously massive and intimate scale of seductive persuasion. This attempt at normalising falsehood does not intend to desensitise, but to stimulate the same impulse as truth, which is to incite action — but action built from a tissue of lies. If we are to believe that the self and truth do hold some continuity within us, which will be all the more tested as non-human intelligences begin to enter into the scripts of daily life, then every form of impact narrative and every step taken toward veracity and hospitality are the counteractions of a counter-we to the imperiously punitive entitlement of manipulative sovereignties. Sontag writes toward the end of *Regarding the Pain of Others,* "To designate a hell is not, of course, to tell us anything about how to extract people from that hell".[5] Our need and our job of self-authorised humanity is to find and tell truths in the belief that the language of our impact narratives is in the name of just laws, whether factual or aspired to, and that we can name in that utterance of equity the ethics of our institutions and how we can deepen what it means to know oneself and to live with others.

Notes

1. Benjamin Moser, *Sontag: Her Life and Work* (New York: Ecco Press, 2019). Susan Sontag, *Regarding the Pain of Others* (New York: Farrar, Straus and Giroux, 2003).

2. Michel Foucault, *Security, Territory, Population: Lectures at the Collège de France, 1977–1978,* ed. Michel Snellart, trans. Graham Burchell (New York: Palgrave Macmillan, 207), 144.

3. Of course, disinformation about Covid-19 goes beyond moral damage into the realm of colossal lethal harm.

4. See their essays in *What about Activism?* ed. Steven Henry Madoff (Berlin: Sternberg Press, 2019).

5. Sontag, "Regarding the Pain of Others", 114.

References

Foucault, Michel. *Security, Territory, Population: Lectures at the Collège de France, 1977–1978.* Edited by Michel Snellart, translated by Graham Burchell. New York: Palgrave Macmillan, 2007.

Madoff, Steven Henry, ed. *What about Activism?* Berlin: Sternberg Press, 2019.

Moser, Benjamin. *Sontag: Her Life and Work.* New York: Ecco Press, 2019.

Sontag, Susan. *Regarding the Pain of Others.* New York: Farrar, Straus and Giroux, 2003.

Triple-Chaser

Forensic Architecture

When US border agents fired tear gas grenades at civilians in November 2018, in Tijuana–San Diego border (Mexico/US), photographs showed that many of those grenades were manufactured by the Safariland Group, one of the world's major manufacturers of so-called "less-lethal munitions". The Safariland Group is owned by Warren B. Kanders, the vice-chair of the board of trustees of the Whitney Museum of American Art.

Whereas the export of military equipment from the US is a matter of public record, the sale and export of tear gas is not. As a result, it is only when images of tear gas canisters appear online that monitoring organisations and the public can know where they have been sold, and who is using them.

But this kind of manual research is laborious, and time-consuming. Automating any part of that process could be hugely beneficial to human rights monitors, and the pursuit of corporate accountability in the global arms trade.

In response to our invitation to the *2019 Whitney Biennial,* and the controversy of Warren B. Kanders's association with the institution, Forensic Architecture began a project to train "computer vision" classifiers to detect Safariland tear gas canisters among the millions of images shared online.

Based on conversations with organisations including the Israeli NGO B'Tselem and the UK-based Omega Research Foundation, we took as our test case a Safariland-manufactured grenade known as the Triple-Chaser.

The task of training a computer vision classifier to identify a particular object usually requires thousands of images of that object. Images of the Triple-Chaser, however, are relatively rare.

To fill the gap, we constructed a digital model of the Triple-Chaser, and located it within thousands of photorealistic "synthetic" environments, recreating the situations in which tear gas canisters are deployed and documented. In this way, "fake" images help us to search for real ones, so that the next time Safariland munitions are used against civilians, we'll know.

In partnership with Praxis Films, we presented the story of this research project as a video investigation, which premiered at the *2019 Whitney Biennial.* At the request of Decolonise This Place, an activist collective leading weeks of protest against

Kanders's connection to the Whitney, Forensic Architecture proved, in April 2019, the presence and use of Safariland products by police during civil unrest in Puerto Rico in 2018. In May 2019, we released our open source investigation into the US munitions manufacturer Sierra Bullets, which exposed the connection between Kanders, Sierra Bullets, and violence against civilian protesters by the Israeli military in Gaza.

On 20 July 2019, Forensic Architecture withdrew from the *2019 Whitney Biennial.* The decision to withdraw, along with seven other artists, was the result of inaction by the Whitney Museum in response to the allegations against Kanders. A few days later, on 25 July 2019, Kanders resigned from the board of trustees of the Whitney Museum of American Art, following protests led by activist group Decolonize This Place. In direct response to his resignation, we rescinded our request to have our work withdrawn, along with several other artists. On 9 June 2020, Kanders announced that he will divest his company of divisions that sell chemical agents, including tear gas, amidst the use of the Triple-Chaser tear gas grenade by police against Black Lives Matter activists across the US.

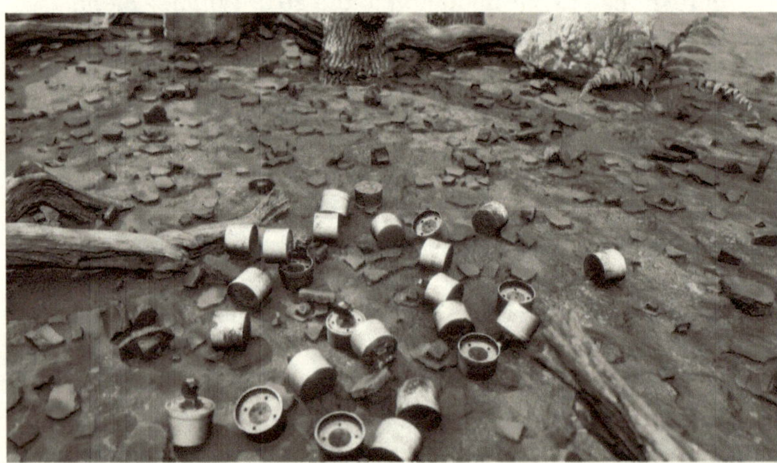

Figure 1. Using the Unreal Engine, Forensic Architecture generated thousands of photorealistic "synthetic" images, situating the Triple-Chaser in approximations of real-world environments (Image: Forensic Architecture, 2019).

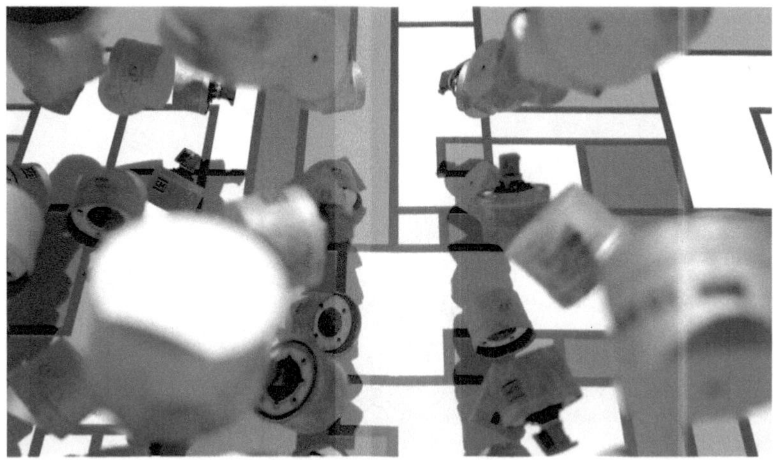

Figures 2–7 & 9. Rendering images of our model against bold, generic patterns, known as "decontextualised images", improves the classifier's ability to identify the grenade. Background image by Patterncooler (Image: Forensic Architecture, 2019).

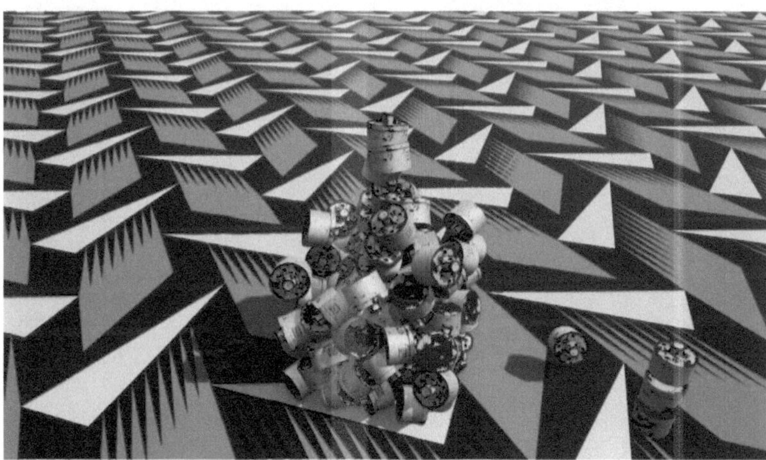

Figure 3.

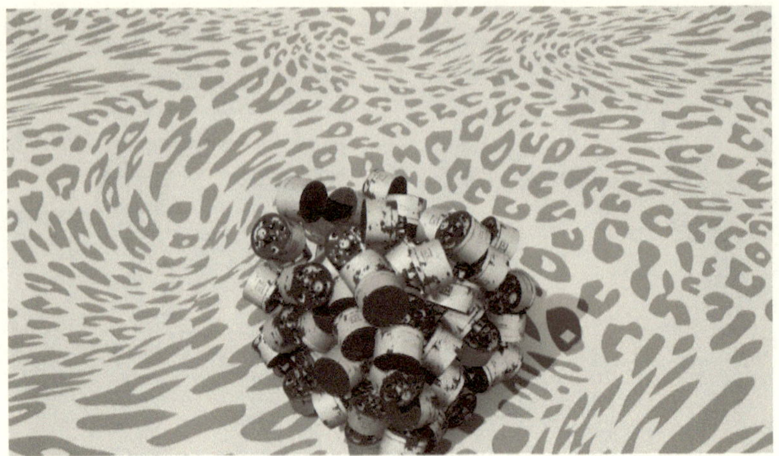

Figure 4.

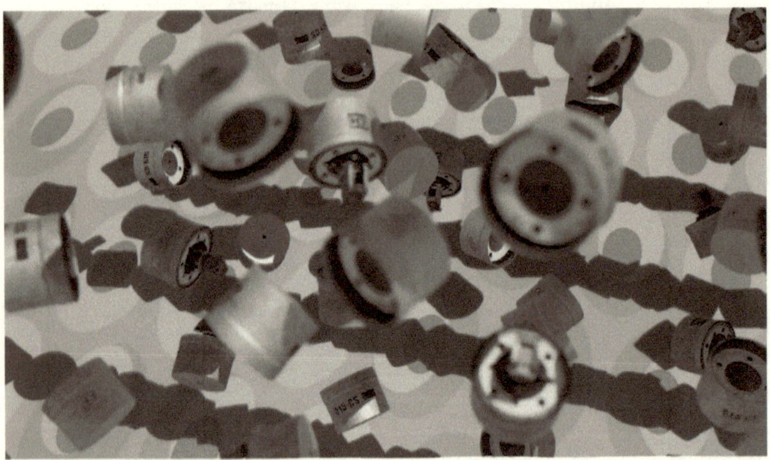

Figure 5.

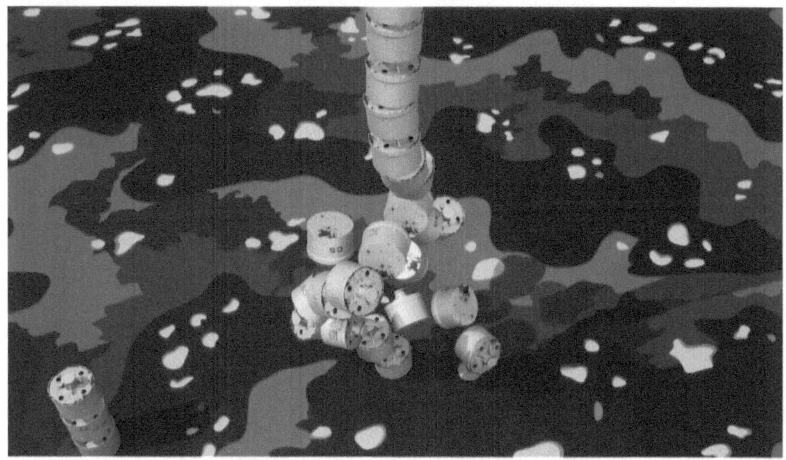

Figure 6.

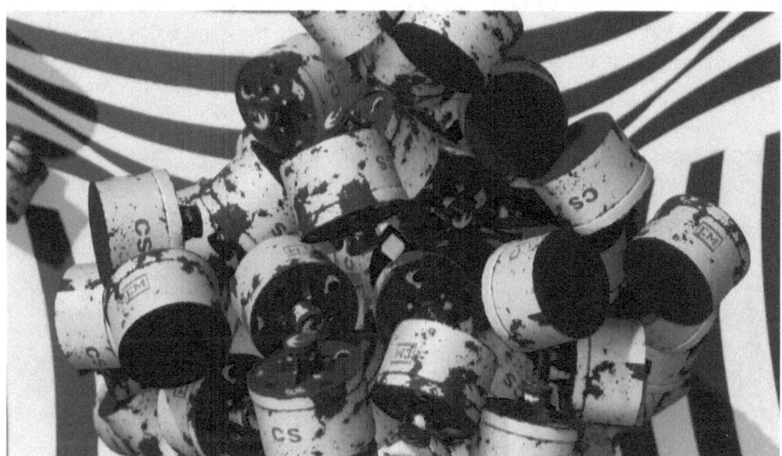

Figure 7.

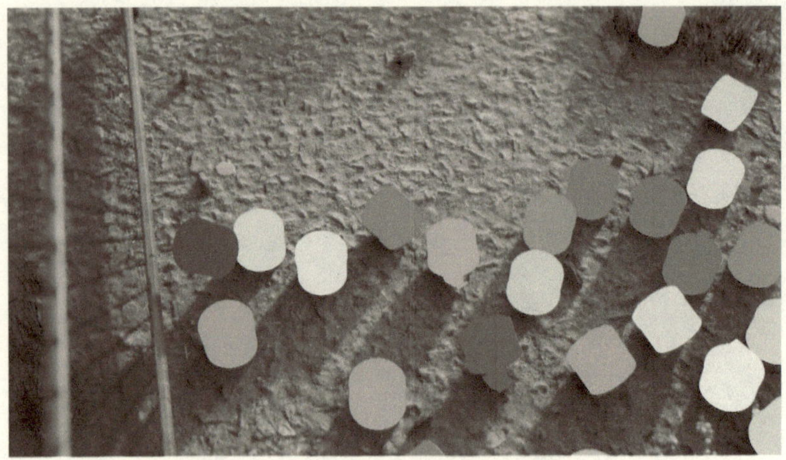

Figure 8. Using the Unreal Engine, Forensic Architecture generated thousands of photorealistic "synthetic" images, situating the Triple-Chaser in approximations of real-world environments. Coloured "masks" tell the classifier where in the image the Triple-Chaser grenade exists (Image: Forensic Architecture, 2019).

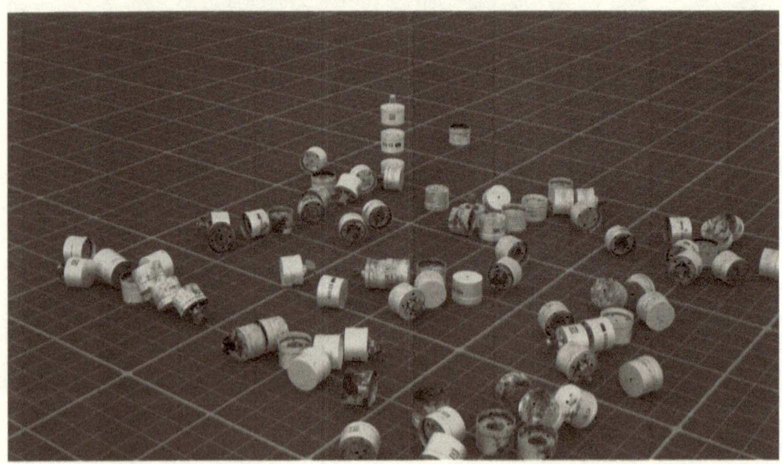

Figure 9.

Post-Anthropogenic Truth Machines

Ramon Bloomberg

The guy said, "You want to know the truth?"

I didn't, but he was going to say it anyway. I watched as he prepared himself to deliver a novelty, or some surprise.

The truth bites

About an hour's drive from the Afghanistan–Pakistan border, on an old airstrip not far from the town of Khost, is a place the Americans call Camp Chapman. The camp is a cluster of untidy buildings at the side of a runway, inhabited by CIA officers and their friends and allies. At the end of December, in that part of Afghanistan, astronomical twilight kicks in at around 6PM. So, at 5PM on 30 December 2009, when Humam Khalil Abu-Mulal Al-Balawi was driven through the security barriers of the base, the CIA officers, their friends, and allies, stepping out to greet their guest, would have had a pretty nice sunset view of the low mountains to the south.

To the CIA officers, Al-Balawi was a Kuwaiti-born double agent, currently operating out of the Federally Administered Tribal Areas (FATA) region of Pakistan. He had been turned in by the Jordanian intelligence services and handed over to the CIA on a silver platter, like a gift. Prior to his alleged turning, Al-Balawi had been a blogger and website personality. The problem was that he had blogged about Jihad and Islam in a positive and convincing way, so he'd attracted the wrong kind of attention from the Jordanians.

These CIA officers and their friends and allies weren't at Camp Chapman for the view. They were there to try and figure out who, in the FATA regions of Pakistan, was worth sending a drone to go and kill. That was their job and their purpose. They needed to sub-contract human agents like Al-Balawi, because otherwise they didn't have the slightest idea of what was going on over the border.

There, in the FATA, a loose network of people, whom the CIA labelled "Al-Qaeda", were living in and among the local Pashtuns. In fact, for the Americans, these "Al-Qaeda" people were indistinguishable from the Pashtuns, whose lives and habits, language and culture, dreams and metaphors they found incomprehensible.

Unfortunately for the CIA officers and their collaborators, Al-Balawi was not thinking about helping them track down and kill his friends and allies in Pakistan; he was thinking about killing the CIA officers right there and then in Camp Chapman — taking the fight to the enemy. Al-Bilawi stepped out of the car and exploded a suicide vest, killing himself along with five CIA officers, two CIA contractors, a Jordanian intelligence officer, and the Afghan guy who drove him across the border. In total, nine were killed and six others seriously wounded.

The bomb sewn into Al-Bilawi's suicide vest made a pretty big bang in Camp Chapman that day, but the blow-back swept through the entire American military truth-making apparatus. This event was the largest single loss of life that the CIA had experienced since 1983.[1] All of which made it easier for certain factions within the American intelligence apparatus to argue for a change of tack. Old-fashioned human-based intelligence was on the way out, and other means were on the way in.

This essay traces the production of truth within resourceful military institutions from the anthropogenic traditions of Human Intelligence to new epistemological practices, in which the source of truth is increasingly distanced from the individual human being.[2] As an operational institution with concrete declared goals, it is possible to trace the ways in which the essentialising of truth-making has been challenged by operational concerns. American intelligence institutions are a dynamic laboratory for epistemological practice. Love them or hate them, the US military has been an active-duty force since their invasion of Afghanistan in 2002, and the endless global wars since then. Put bluntly, one truth that the military is continuously chasing is the question of who to kill, or in other words, what is a target. This question is both focused, and more complicated than it may seem, opening up questions of agency and authorship that may be useful to other institutions, regardless of how they may believe themselves to be exceptional and distinct from military practices.[3]

Human Intelligence

It is possible to trace attempts by American military intellectuals to attach a stable definition to the term "intelligence".[4]

Most formalised approaches during the post-war period have referred almost exclusively to the collection and production of *knowledge.*[5] This emphasis on knowledge seems to ignore an important temporal aspect of the intelligence concept.

For Alain Badiou, knowledge is what transmits, what is repeated. Truth is something *new.*[6] Bringing in a temporal frame enables us to gain traction on the concept of truth, or at least on one level. Intelligence is considered, here, to be the search for new truths that might then be distributed as institutional knowledge. Intelligence about this or that subject or theme is always going to be the latest stuff, not last-year's stuff. Last year's stuff isn't intelligence, it's just last year's stuff.

The framework for producing institutional knowledge that involves running a double agent is known as Human Intelligence. Human Intelligence has historically been considered the most effective means of generating intelligence.[7] People speak and act, and there are recordings of their speech and acts such as books, films, emails, text messages, telephone conversations, papers in folders, and photographs.

Signals Intelligence has traditionally concentrated upon the interception and decoding of messages, with a view to unlocking and capturing the content of the message. Thus, in a sense, a form of Human Intelligence. That is to say, even in Signals Intelligence, the content, novelty, and surprise of the intel is extracted at the scale and temporality of an individual human being.[8]

In an article from the journal *Military and Strategic Affairs,* an output of The Institute for National Security Studies in Tel Aviv,[9] Bradley Lewis decries the death of Human Intelligence (HUMINT) from the 1960s to the present,[10] in the increasing reliance on what he terms "the use of technology".[11] Lewis writes that built-in limitations exist on the capacity of any intelligence organisation to field qualified personnel for handling agents. Typically, only five percent of an organisation is capable of being dedicated to building a human-source network, thus limiting the possible scale of these connections.[12] The work is dangerous, reprisal killings of suspected agents being common, along with the risk to handlers.[13]

Lewis suggests that HUMINT-intensive methods for intelligence extraction such as torture and interrogation are being formally disavowed, because institutional faith has been displaced towards technical means. This is the case not just in terms of operational style, or habit, but has been codified into law. He writes: "As technology has grown and functionality has improved, the need for HUMINT, as determined by current policy, has increasingly diminished. The Obama administration determined that the use of HUMINT in many forms is a punishable offence".[14] The regulation of HUMINT is legally binding as of Intelligence Community Directive Number 304, dated March 2008.[15]

Al-Balawi, the double agent, was employed by the CIA in order to generate drone targets. This was when drone strikes were still being made against known individuals with names. After the suicide bomb at Camp Chapman in 2009, the Human Intelligence framework was formally disavowed by the Americans. New methods of producing a target were already being employed in places like Iraq. Human Intelligence was dismissed as difficult and messy, if not thoroughly untrustworthy.

Action and the political sphere

Another way of describing Human Intelligence is as an anthropogenic engine for generating the novel, the unexpected, and the surprising into institutional knowledge flows.

The individual human being is front and centre — the source of the new. As such, these practices are congruent with the conceptual system that Hannah Arendt describes as a *political sphere,* in which human speech and action are the source of the new and unexpected, bringing the individual and plurality into a mutually constitutive relation.[16]

This abstract configuration of the political sphere in Arendt is a *space of appearance,* in which individual human action is a primordial revelation of both plurality and distinction. There is no distinction without plurality, which can never be a simple multiplication of the distinct. The *political* privileges freedom, understood as the possibility of beginning, of initiative.[17]

This is key, because of how, in Arendt, the possibility of a new truth is inextricable from the public appearance and natality of an individual human being. To appear in public is, for Arendt, a second birth. After the biological birth into the social household (*oikos*), the second birth introduces a political being into the space of appearance, within the sphere of politics.

Speech and action together form the structuring initiatives that distinguish individual humanity qua humans. Humans appear to each other as human and in distinction to physical objects, which merely exist. Agency, in Arendt's political sphere, is securely located at the address proper to an individual human body.

Actionable history and time

Suhail Malik suggests that Arendt's formulation locates *an emergence of actionable history* — a configuration that distinguishes the temporality of modernity from that of a previous Euro-Christian eschatology. Roughly, from the mid-seventeenth century, a modern framework emerges that dispenses with the Last Judgement as the horizon and limit of futurity. In its place is human action.[18] As noted above, the unique capacity of the individual human to initiate action is understood as the condition of an unexpected and limitless future, now liberated from the certain arrival of doomsday characterising the temporal horizon of Euro-Christian pre-modernity.

The organisation of modern anthropogenic futurity was fortified by a temporal scale of *development and progress,* inextricably linked to European expansion and colonisation. From early modernity, the globe and the global produced not only the disinhibition necessary for active globalising, but an awareness of simultaneous location space. That is to say, the globe, as a device, allowed people to construct, for the first time, mental projections of themselves and others in a new kind of spatial construct.[19]

The spatial simultaneity of the globe was temporally organised along a spectrum of development and progress.

Spaces of appearance and ontologies with a lowercase "o"

In the modern framework, anthropogenic production of the new is extensible, from the individual human being to the material forms that human action may crystallise. One highly *developed* form of Arendtian action has been the literary work. And, not only has institutional intelligence production generated a literary form, the espionage novel, but literary action itself has often been the subject of this literature.

In John Le Carré's famous series of spy novels,[20] the retired spy George Smiley possesses the ability to wade through reams of paper with an inhuman (lizard-like) concentration. Smiley methodically thumbs through documents, organising, making minute pencil marks. His is a cunning of tabulation and sorting. The patterns and temporalities of the Circus (moniker for the foreign intelligence office), are rendered in the particular grammar found in the dossier. The dossier is composed of literary work (written documents), and photographs (of installations and people). Smiley, as an intelligence agent, possesses a literary cunning. It is significant that besides being a spy he is an expert in Baroque German literature, for his field of activity is primarily East Germany.

Smiley's agency derives firstly from his being human and, therefore, a valid modern source of newness, or truth. Secondly, Smiley enjoys a highly developed ability to immerse himself in literature — one of the ways in which human action may be materialised, extended from the human body. Literature, and its grammar or syntax, share the temporalities and conditions associated with human experience. To be inside of a dossier, or book, or speech act, is to exist in the interior of what computer scientists refer to as an "ontology", with a lowercase "o".[21] Roughly speaking, a term that aims to describe the practical application of the philosophical notion of Ontology (written with an uppercase "O") as that which seeks to account for existence.

In literature, this lowercase "o" is extensible all the way to the uppercase "O". The literature generated by Human Intelligence operations has an *external referent* at a planetary scale and a planetary temporality — that is to say accessible and sympathetic to the circadian rhythms and the cosmos, more generally. Any human being has the potential to enter into the literary world; it's really just a matter of *developing* the necessary capacities that are innate to the *anthropos.*

In *The Wretched of the Earth* (1961), Frantz Fanon discusses the post-colonial development of the Algerian people, a "mass of starving illiterates".[22] Fanon writes that "one of the greatest services that the Algerian revolution will have rendered to the intellectuals of Algeria will be to have placed them in contact with the people, to have allowed them to see the extreme, ineffable poverty of the people, at the same time allowing them to watch the awakening of the people's intelligence and the onward progress of their consciousness".[23] This is a quintessentially modern configuration of progress, development, and literature. Fanon describes spatially simultaneous human-beings who are temporally ordered in terms of their developmental progress.

Time's up

Malik suggests that Arendtian modernity, coincidental with anthropogenic futurity (humans make the future), is in the process of being displaced by the emergence of a new temporal order, more or less defining the current period, although of course not entirely replacing it, but layered in and around vestiges of modernity.

This transformation is conditioned by, and contextualised within what he terms "large-scale integrated complex societies" (LaSICS): "Whereas modernity is structured by a horizon of expectation, a new future to come that is distinct from the present, what is by comparison distinct to LaSICS is that the futurity of the new is their functional condition, the operational premise for their technical, material, and symbolic organisation and development".[24]

In other words, the new temporal order — future comes before the present — is not a consequence of globalised and computational integration, but an operational condition for this kind of large-scale integration to occur in the first place. So, we're just at the beginning.

Arendt's modern horizon of expectation is incarnated by the individual human potential for action, understood as initiative. If, in modernity, past, present, and future were organised according to a hierarchal temporality of *development* and *progress,* this is no longer the case. Now, the present is actualised by a speculative accounting, only possible by making the future its premise.[25]

This formulation is less alien once we consider the manner in which the operation of large-scale integrated complex societies is grounded in financialised systems of credit and generalised speculative anticipation. As Malik points out, this is readily apparent in activities like insurance, banking, healthcare, energy, agriculture, and logistics.[26]

The apocalyptic discussions around climate change demonstrate the difficulties that people have adapting to new temporalities. In much of the climate change discourse, the future is so highly valued that valuation of the present suffers as a consequence. This upends a normative formula in which the future is discounted because the present is more highly valued.[27]

Anticipatory futurity has also become the basis for evaluating and producing the present within the intelligence institutions of large state bureaucracies.

Activity-Based Intelligence

If the literary exemplifies the relations between human action and its material, mediatic crystallisation, drawing a line from the human-being through physical objects such as documents, poems, cassette tapes, sculptures, paintings, and novels, it was possible, in modernity, to find truth in these places.

But, the truth does not seem to be emerging from the usual places anymore. Or, at least, truthful revelation from individual human action is decreasing. It's as if the well has run dry.

What is the future of the new?

It has been suggested, that the US military's search for new sources of truth was focused by problems encountered fighting the Iraq insurgency during 2004–06.[28]

Amidst the insurrection of both Shiite and Sunni Iraqis, the Americans struggled to identify legitimate targets from within the general population. The matter was viewed as a sorting problem. The difficulty lay in finding the means to filter, or isolate Iraqis who might be legitimately killed or captured, from those whose death or capture would be seen as illegitimate. This stemmed from the fact that Iraqi combatants were visually indistinct from non-combatants and intermingled with them.[29]

What the military lacked in its operational capacity was the ability to distinguish targets (truths) by identifying markers such as uniforms and military hardware.

Workarounds were improvised using the new technical systems and objects that network-centric warfare doctrine had delivered. Drones, mobile phone data, video cameras, technical networks, GPS, spreadsheets, and meta-data.

Geo-spatial analysts from the National Geospatial-Intelligence Agency (NGA) established databases with geo-referenced data, gleaned from multiple sources, and began to deliver *adversary* locations.

These geo-locations became areas of *interest,* with the people found in them becoming *persons of interest* by association. Analysts looking at full motion video (FMV) from drones were the first to produce *pattern-of-life analysis.* From this matrix — or "Multi-INT fusion"[30] — geo-locations were sorted along a spectrum of probable risk.

The crucial point here is that, while previous counter-insurgency tactics had aimed at targeting insurgents, the new methodologies, in journalist Gareth Porter's words, targeted "phone numbers not people".[31] Epistemologically, one result of this new truth-making ensemble is that individual human beings were no longer the *a priori* basis of target production.

These new practices contributed to the ascent of the Activity-Based Intelligence (ABI) doctrine within the circuits of military truth-making.

Weak signatures

In ABI, identifying markers such as military uniforms and military hardware are known as strong signatures. ABI seeks out what are referred to as weak signatures. The needle-in-the-haystack seems like a pretty good metaphor to describe hunting insurgents within a general population. Except it isn't; the needle has a strong signature, and so does hay.

But, what happens when you are looking for a stalk of hay in a haystack?

ABI begins with the collection of sensed data, employed to record activities and transactions, indexed to their geo-reference over time. Examples range from the physical movement of a person, motor vehicle, train, or animal to other sensed phenomena such as electronic messages, telephone conversations, or the heat emissions of electrical devices operating over time.

Activities and transactions are categories of events; occurrences within time and space. Events are sensed and stored as data. As Gaston Bachelard notes, each sensor is an instrument capable of approaching phenomena at a specific order of magnitude.[32] One implication is that, at the same time, the sensor neglects the orders of magnitude not being sensed.

Thus, a sensor harvesting electronic emissions is not capable of perceiving that a person named Jim from a town called Trouble is carrying a mobile phone on Main Street. The sensor perceives a signal emitted of x strength. Any further knowledge must be extrapolated by correlating data from other sources.

One way in which ABI purports to find hay in a haystack is through the establishment of historical patterns. Pattern of Life (POL) mappings are statements of a baseline norm, against which deviations are registered as anomalies.

Continuing with the hay-in-a-haystack metaphor, practitioners might take the stack as a three-dimensional cube and measure its humidity over, say, a month under given conditions with a hygrometer. This would return sufficient data to draw a POL, or Pattern of Humidity (POH) — mapping of the haystack. Under persistent hygrometric surveillance, changes in the humidity levels above or below a threshold, defined by an arbitrary decision based upon the POH map, will be flagged as anomalies, i.e. as needles.

This is not very distant from the way that ABI epistemologies understand human settlements under remote controlled occupation. Which is to say that, first of all, they don't.

For the hygrometer there is only humidity, not hay. Within data ontologies at this stage in its process, ABI does not understand the object of its inquiry as human or hay, for that matter. Entities are understood in the terms of thresholds of intensity, determined by the constraints imposed by the technical system. It is a fundamental of ABI that judgements or identifications are deferred.

Two metaphors are deployed in distinguishing ABI from other forms of knowledge production: the puzzle versus the mystery.[33]

A puzzle addresses known problems; the analyst knows that there is a piece that will complete the picture, an order of things that just needs to be worked out. In the puzzle metaphor, there was once a complete entity (the past), which has been disrupted (the present), for which finding the final piece will resolve the problem (the future). The investigative activity runs along a defined timeline, the horizon of which is completion. This corresponds to the modern temporal framework.

In the mystery metaphor, it is not clear that there is even a puzzle to solve. The analyst collects data and makes connections, always with an eye to the possibility of being surprised, to the possibility of finding a puzzle that can be solved.

This is a radically open-ended mobilisation of resources. According to military intelligence intellectuals, there is no *a priori* entity inciting the investigation (truth-making activity). There is a stance of non-judgemental openness to the world: "The analyst is not cued or focused on a specific target, but rather is informed by the data as it is being presented".[34]

Yet, this is misleading, as there must be an explanatory basis for mounting an investigation and mobilising a knowledge-seeking apparatus, even one that remains operational for an extended, indefinite period. The *a priori* entity worthy of this investigation of a mystery seeking a puzzle is the *Unknown-Unknown.*[35]

Unk-Unk

Put simply, an Unknown-Unknown is a state of ignorance at a specific point in time for all members of the organisation.[36]

This formulation derives from high-risk research and development cultures, like the aerospace industry.[37] The disposition of an organisation to recognise the Unknown-Unknown, as a condition, is encouraged as a stimulant to a speculative line of thinking.[38] The Unknown-Unknown is not equivalent to a risk, which is a known unknown, rather, it is the equivalent of a surprise.

That is to say, just because an activity or transaction is phenomenologically unavailable at the present time, does not validate any knowledge around its possible (non-)existence. The investigator must always be open to the discovery of a new problem, one that she did not previously know that she had to solve. With the Unknown-Unknown, it is not necessary to develop a probable-cause rationale for a surveillance operation, because all prior suppositions are considered deceitful.

Concurrently, the figure and formulation of the Unknown-Unknown has re-oriented the disposition of the American military intelligence-seeking apparatus to the determined elision of presupposed knowledge. In addition, and regardless of cause, what begins to be apparent is a methodological shift away from an anthropocentric scale to orders of magnitude more proper to activity and transaction and the instruments that are most likely to record their traces.

Activity without action

Arendtian action is an "initiative from which no human being can refrain and still be human".[39] Action is intimately tied to the individual human being. Activity is not.

Anthropogenic initiative provides a secure location for agency, the individual human being. This is a conceptual foundation for constructs such as human rights. The idea that individual human beings have the innate capacity to deliver new and unexpected surprise and truth to the world, more or less equally, underpins the modern political structure.

In the modern framework exemplified by Human Intelligence, both the future and the past may be conditioned in the present by anthropogenic initiative — the agentic location of the arbitrary and unexpected *new.* ABI institutionalises what Malik terms the Speculative Time Complex (STC), a disposition in which the future is both a prior and subsequent condition for the present: "The STC is the schematic configuration of the unknown future as the operational prior condition of the present".[40]

Activity is not equivalent to action.

Firstly, action and activity exist at different orders of magnitude. If action is tied into the individual human being, activity receives its inputs from a multitude of what moderns would consider to be "actions". ABI collates multiple points of action, crucially ignoring (procedurally neglecting) the source of action for an appreciation of the correlation between multiple activities and transactions, time, and space.

But, activity is not apprehended at the order of magnitude proper to population either. It is not a zooming out to the macro. Activity is not even in the same spatial register. Rather, the apprehension of activity eschews topographical time and space for the continuum of dynamic topological matrices.

Narrative drama theory might be able to help make the point. For theorists such as Konstantin Sergeievich Stanislavski, Sanford Meisner, Robert Bresson, and David Mamet, the production of *character* is only a function of the accumulation of action. The idea of character, that is to say, the appreciation of an individual human being, distinct from others, is produced in the mind of an audience through the steady accretion of actions over time. Note that the accumulation of action into character is concentrated in the agentic body of an individual human being.

By contrast, ABI accumulates a matrix of activities and transactions to produce historical patterns, decontextualised and agnostic on what the source of activity is — human, machine, animal, etc. Rather than addressing an individual or mass of individuals (as in the concept of population), the new forms of truth-making are addressed to an "infra and supra-individual statistical body",[41] in which truth is always already present as a "memory of the future",[42] without recourse to the physical and temporal frame of an individual human being.

Becoming truth machines

While technologically adept and resourceful institutions such as the US intelligence apparatus, NASA, high-end hedge funds, and insurance companies may derive truth statements from infra- and supra-individual statistical bodies and actuarial tables, institutions lacking these means are left in a post-truth condition. It's not that truth has disappeared from the world, but that access to the production of truth has been displaced.

This is the true meaning of the expression post-truth.

While doing away with the potency of the modern liberal subject, the new truth-making techniques at least constitute an alternative. However, it's an expensive machine to build.

In the modern era, the anthropogenic truth machine was open-source /open-access to human beings. Through a combination of *logos* and biology, the anthropogenic truth machine provided institutions and people alike with the means for producing truth statements. It was simply a matter of pointing the finger at an individual human being, present or latent. Artists and Authors, Geniuses, Generals, and Mavericks, these modern figures could even be raised on a plinth outside monumental buildings to stand in for a previous epoch's raised crosses.

Who would have imagined that truth-making was in fact a technical arms race?

Notes

1. See an open letter to the CIA from President Obama, https://www.whitehouse.gov/the-press-office/message-president-cia-workforce.

2. By "resourceful", I mean possessing the means to mobilise copious resources such as large budgets, materiel, and manpower.

3. For recent discussions on the so-called "dual use" transfer of technologies and processes between military and civil contexts, see Alison Howell, "Neuroscience and War: Human Enhancement, Soldier Rehabilitation, and the Ethical Limits of Dual-Use Frameworks", *Millennium,* Vol. 45, No. 2 (January 2017): 133–150; Dagmar Rychnovská, "Governing Dual-Use Knowledge: From the Politics of Responsible Science to the Ethicalization of Security", *Security Dialogue,* Vol. 47, No. 4 (August 2016): 310–328; and the work of Paul Virilio, more generally.

4. For example, Michael Warner, "Wanted: A Definition of 'Intelligence'", Central Intelligence Agency, last modified 27 June 2008, https://www.cia.gov/library/center-for-the-study-of-intelligence/csi-publications/csi-studies/studies/vol46no3/article02.

5. For example, Vernon A. Walters, *Silent Missions* (Garden City, NY: Doubleday, 1978), 621; and Martin T. Bimfort, "A Definition of Intelligence", Studies in Intelligence, Vol. 2 (Fall 1958): 75–78, https://www.cia.gov/library/readingroom/docs/DOC_0000606548.pdf.

6. Alain Badiou and Justin Clemens, *Infinite Thought: Truth and the Return of Philosophy,* trans. Oliver Feltham (London: Bloomsbury, 2014), 61.

7. Saxby Chambliss, "We Have Not Correctly Framed the Debate on Intelligence Reform", *Parameters,* Vol. 35, No. 1 (Spring 2005): 6; Jennifer Sims, "Transforming U.S. Espionage: A Contrarian's Approach", *Georgetown Journal of International Affairs,* Vol. 6, No. 1 (Winter/Spring 2005): 54; and Rand C. Lewis, "Espionage and the War on Terrorism:

Investigating U.S. Efforts", *The Brown Journal of World Affairs,* Vol. 11, No. 1 (Summer/Fall 2004): 175.

8. I am making a distinction here between classic SIGINT and more recent interception methods that are aimed at meta-data and the like. Those methodologies are the subject of the second part of this essay.

9. Lewis, "The Death of Human Intelligence: How Human Intelligence Has Been Minimized Since the 1960s".

10. Lewis, "The Death of Human Intelligence: How Human Intelligence Has Been Minimized Since the 1960s", 77. Lewis cites the FBI definition of HUMINT: "Human Intelligence (HUMINT) is the collection of information from human sources. The collection may be done openly, as when FBI agents interview witnesses or suspects, or it may be done through clandestine or covert means (espionage). Within the United States, HUMINT collection is the FBI's responsibility".

11. Lewis, "The Death of Human Intelligence: How Human Intelligence Has Been Minimized Since the 1960s", 80.

12. Lewis, "The Death of Human Intelligence: How Human Intelligence Has Been Minimized Since the 1960s", 77.

13. "Fears of reprisal are palpable and their consequences are dire. In Afghanistan alone, the United Nations observed [...] 462 assassinations in 2010 in reprisal for cooperating with the coalition according to their records, double the number from the previous year. The figures may not include many killings in remote areas, like the mass beheading, because fearful villagers never reported them". Ray Rivera, Sharifullah Sahak, and Eric Schmitt, "Militants Turn to Death Squads in Afghanistan", *New York Times,* 28 November 2011. Quoted in Lewis, 2016.

14. Lewis, "The Death of Human Intelligence: How Human Intelligence Has Been Minimized Since the 1960s", 84.

15. "The DNI is committed to ensuring that HUMINT activities are executed in a prioritized, coordinated, integrated, and professional manner; that USG elements engaged in the collection of intelligence through HUMINT activities, counterintelligence activities, or activities that involve the use of clandestine methods are coordinated and deconflicted with IC HUMINT activities; that HUMINT practitioners use core common standards; and that there is transparency into HUMINT support capabilities to allow all IC elements to benefit from technical or other advances". *Intelligence Community Directive (ICD),* #304, Human Intelligence, 6 March 2008, amended 9 July 9 2009. Quoted in Lewis, 2016, 78.

16. Arendt, "The Human Condition", 26.

17. Arendt, "The Human Condition", 178.

18. Suhail Malik, "Contra-Contemporary", In *The Future of the New: Artistic Innovation in Times of Social Acceleration,* ed. Thijs Lijster (Amsterdam: Valiz, 2018), 252.

19. Disinhibition is a fundamental concept in Peter Sloterdijk's lexicon: "The concept of disinhibition, without which no convincing theory of modernity is possible, gathers together the motives that drive us to intervene in the imperfect and disagreeable". Peter Sloterdijk and Wieland Hoban, *In the World Interior of Capital: Towards a Philosophical Theory of Globalization* (Cambridge, UK: Polity, 2015), 19.

20. John Le Carré, The spy who came in from the cold (London: Pan Books, 1965). This is the first novel in which Smiley appears as a central character.

21. "An ontology is an explicit specification of a conceptualization. The term is borrowed from philosophy, where an ontology is a systematic account of Existence. For knowledge-based systems, what 'exists' is exactly that which can be represented". Thomas R. Gruber, "A translation approach to portable ontology specifications", *Knowledge Acquisition* 5.2 (1993): 199.

22. Frantz Fanon, *The Wretched of the Earth,* trans. Richard Philcox (New York: Grove Press, 2004), 130.

23. Fanon, "The Wretched of the Earth", 130.

24. Malik, "ContraContemporary", 13.

25. Malik, "ContraContemporary", 14.

26. Malik, "ContraContemporary", 14.

27. For an interesting discussion of the psychology of delay discounting, see Trishala Parthasarathi et al., "The Vivid Present: Visualization Abilities Are Associated with Steep Discounting of Future Rewards", *Frontiers in Psychology* 8:289 (June 2017), https://doi.org/10.3389fpsyg.2017.00289.

28. Patrick Biltgen and Stephen Ryan, *Activity-Based Intelligence: Principals and Applications* (London: Artech House 2015), 24.

29. This is not a new problem. For an historical precedent, see the CIA's Phoenix programme during the Vietnam war. The solution then, measured in body-count statistics, resulted in the murder of at least 8,000 unarmed civilians.

30. Biltgen and Ryan, *Activity-Based Intelligence,* 24.

31. Gareth Porter, "How McChrystal and Petraeus Built an Indiscriminate 'Killing Machine'", *Truthout,* 26 September 2011, https://www.truth-out.org/news/item/3588:how-mcchrystal-and-petraeus-built-an-indiscriminate-killing-machine.

32. Gaston Bachelard, *The formation of the scientific mind,* trans. Mary McAllester Jones (Manchester, UK: Clinamen Press, 2002 [1938]), 222.

33. The origins of this concept are worth quoting in full. From Gregory Treverton's 2007 Smithsonian article: "There's a reason millions of people try to solve crossword puzzles each day. Amid the well-ordered combat between a puzzler's mind and the blank boxes waiting to be filled, there is satisfaction along with frustration. Even when you can't find the right answer, you know it exists. Puzzles can be solved; they have answers. But a mystery offers no such comfort. It poses a question that has no definitive answer because the answer is contingent; it depends on a future interaction of many factors, known and unknown. A mystery cannot be answered; it can only be framed, by identifying the critical factors and applying some sense of how they have interacted in the past and might interact in the future. A mystery is an attempt to define ambiguities. Puzzles may be more satisfying, but the world increasingly offers us mysteries. Treating them as puzzles is like trying to solve the unsolvable — an impossible challenge. But approaching them as mysteries may make us more comfortable with the uncertainties of our age". Gregory F. Treverton, "Risks and Riddles", *Smithsonian Magazine,* 1 June 2007, https://www.smithsonianmag.com/history/risks-and-riddles-154744750.

34. Treverton, "Risks and Riddles".

35. Treverton, "Risks and Riddles".

36. Roberts, "Organizational ignorance" 107.

37. Roberts, "Organizational ignorance", 107.

38. Roberts, "Organizational ignorance", 107.

39. Arendt, "The Human Condition", 176.

40. Malik, "ContraContemporary", 14.

41 Rouvroy, "The end(s) of critique", 157.

42. Rouvroy, "The end(s) of critique", 157.

References

Arendt, Hannah. *The Human Condition.* Chicago: University of Chicago Press, 2012.

Bachelard, Gaston. *The formation of the scientific mind.* Translated by Mary McAllester Jones. Manchester, UK: Clinamen Press, 2002 [1938].

Badiou, Alain, and Justin Clemens. *Infinite Thought: Truth and the Return of Philosophy.* Translated by Oliver Feltham. London: Bloomsbury, 2014.

Biltgen, Patrick, and Stephen Ryan. *Activity-Based Intelligence: Principals and Applications.* London: Artech House, 2015.

Bimfort, Martin T. "A Definition of Intelligence". *Studies in Intelligence,* Vol. 2 (Fall 1958): 75–78. https://www.cia.gov/library/readingroom/docs/DOC_0000606548.pdf.

Chamayou, Grégoire. "Patterns of Life: A Very Short History of Schematic Bodies". *The Funambulist Papers* 57 (2016). https://thefunambulist.net/history/the-funambulist-papers-57-schematic-bodies-notes-on-a-patterns-genealogy-by-gregoire-chamayou.

Chambliss, Saxby. "We Have Not Correctly Framed the Debate on Intelligence Reform". *Parameters,* Vol. 35, No. 1 (Spring 2005): 5–13.

Fanon, Frantz. *The Wretched of the Earth.* Translated by Richard Philcox. New York: Grove Press, 2004.

Gruber, Thomas R. "A translation approach to portable ontology specifications". *Knowledge Acquisition* 5.2 (1993): 199–220.

Hildebrandt, Mireille, and De Vries, Katja, eds., *Privacy, Due Process and the Computational Turn.* London: Routledge, 2013.

Howell, Alison. "Neuroscience and War: Human Enhancement, Soldier Rehabilitation, and the Ethical Limits of Dual-Use Frameworks". *Millennium,* Vol. 45, No. 2 (January 2017): 133–150.

Le Carré, John. *The spy who came in from the cold.* London: Pan Books, 1965.

Lewis, Bradley A. "The Death of Human Intelligence: How Human Intelligence Has Been Minimized Since the 1960s". *Military and Strategic Affairs,* Vol. 8, No. 1 (July 2016): 75–90.

Lewis, Rand C. "Espionage and the War on Terrorism: Investigating U.S. Efforts". *The Brown Journal of World Affairs,* Vol. 11, No. 1 (Summer/Fall 2004): 175–182.

Lijster, Thijs, ed. *The Future of the New: Artistic Innovation in Times of Social Acceleration.* Amsterdam: Valiz, 2018.

Parthasarathi, T., Mairead H. McConnell, Jeffrey Leury, and Joseph W. Kable. "The Vivid Present: Visualization Abilities Are Associated with Steep Discounting of Future Rewards". *Frontiers in Psychology* 8:289 (June 2017). https://doi.org/10.3389/fpsyg.2017.00289.

Porter, Gareth. "How McChrystal and Petraeus Built an Indiscriminate 'Killing Machine'". *Truthout,* 26 September 2011. https://www.truth-out.org/news/item/3588:how-mcchrystal-and-petraeus-built-an-indiscriminate-killing-machine.

Roberts, Joanne. "Organizational ignorance: Towards a managerial perspective on the unknown". *Management Learning,* Vol. 44, No. 3, (July 2013): 127–146.

Rouvroy, Antoinette. "The end(s) of critique: data-behaviourism vs. due-process". In *Privacy, Due Process and the Computational Turn.* Edited By Mireille Hildebrandt and Katja De Vries. London: Routledge, 2013.

Rychnovská, Dagmar. "Governing Dual-Use Knowledge: From the Politics of Responsible Science to the Ethicalization of Security". *Security Dialogue,* Vol. 47, No. 4 (August 2016): 310–328.

Sims, Jennifer. "Transforming U.S. Espionage: A Contrarian's Approach". *Georgetown Journal of International Affairs,* Vol. 6, No. 1 (Winter/Spring 2005): 53–59.

Sloterdijk, Peter, and Wieland Hoban. *In the World Interior of Capital: Towards a Philosophical Theory of Globalization.* Cambridge, UK: Polity, 2015.

Treverton, Gregory F. "Risks and Riddles". *Smithsonian Magazine,* 1 June 2007. https://www.smithsonianmag.com/history/risks-and-riddles-154744750.

Walters, Vernon A. *Silent Missions.* Garden City, NY: Doubleday, 1978.

Warner, Michael. "Wanted: A Definition of 'Intelligence'". Central Intelligence Agency. Last modified 27 June 2008. https://www.cia.gov/library/center-for-the-study-of-intelligence/csi-publications/csi-studies/studies/vol46no3/article02.

Isuma, the Planetary, Listening and the Reinvention of the Counterpublic Sphere

Christine Ross

This essay is part of a larger research project examining contemporary art's response to the migration crisis and its relatedness to climate change. The research investigates artistic practices that show twenty-first-century migration to be amongst the most imperiled conditions of our times — a condition that turns exodus into a process of marginalisation or elimination of a significant segment of humanity.[1] Contesting that process, artistic practices uphold the requirement to rethink coexistence — the state, awareness, and practice of existing interdependently — not so much a living-together as an inexorable relation between humans, as well as between humans and nonhumans, in urgent need of repair, reciprocity, and looping. Coexistence is also a modality by which art reinvents itself. That reinvention is based on the recognition that migration has now become a "crisis", with this pivotal specification: while the West and the Global North have been describing mass-migration as a crisis due to our alleged inability to integrate migrants, the crisis is in fact elsewhere — it is, more specifically, in the strategies used by state authorities that materialise migrants as a threat to be contained. These strategies have to do with the constant re-categorisation of migrants — as refugees, asylum-seekers, but also war refugees, economic migrants, unauthorised, undocumented, or clandestine migrants — so as to, in effect, delimit, reduce and manage the number of admissible migrants, as well as the escalation of border control policies and militarisation.[2] As anthropologist Nicholas De Genova observes:

> *[i]n the face of the resultant proliferation of alternating and seemingly interchangeable discourses of migrant or refugee crisis, the primary question that must be asked, repeatedly, is: Whose crisis? The naming of this crisis as such thus operates precisely as a device for the authorisation of exceptional or emergency governmental measures toward*

> *the ends of enhanced and expanded border enforcement and immigration policing.*[3]

Questioning that particular social construct, what sociologist Craig Calhoun has called the "emergency imaginary", my understanding of the migration crisis refers *both* to the planetary increase of displaced people since 2011 — migrants displaced by conflict, economic, or any other precarious condition such as global warming and the overexploitation of land, or out of fear of persecution — and the implementation, at least in the West, of national xenophobic immigration policies and border systems that preselect some migrants over others to block as much as possible the movement of the "undesirables".[4]

Of special relevance to this essay is cyberbalkanization — the splintering of the internet community into sub-publics with specific interests, whose division is most likely to be reinforced along regional or national lines and whose modus operandi lies in the systematic avoidance of viewpoints that contradict a given sub-public's belief system. Cyberbalkanization amplifies the waning of a dialectical understanding or truth about what constitutes migration in the twenty-first century and why it has indeed become a crisis.[5] If truth is shared, its sharedness has somewhat shrunk only to reach and reconfirm isolated digital tribes — a segregation that weakens the possibility of recognising other perspectives or establishing a common ground between publics around facts. Belonging to such digital communities has, in itself, become problematic insofar as it is easily unknown by the users themselves, caught as they are in the information streams oriented by the new global media giants, including Google, Facebook, and Twitter. What is art to do; what truth can it distribute in a historical present conditioned by post-factuality? One possible resistance to cyberbalkanization — a form of resistance I will be investigating here — is to expand the publicness of art, and to support art and art institutions that connect different (counter)publics so that they may contest one another and enlarge their worldviews in light of other worldviews, while also making manifest what post-factuality tends to deny: a worldview is always necessarily situated. A worldview's truth value is a perspective, never the ultimate truth. Truth — if we can still use that term — comes from the ongoing relationality of perspectives, on a planetary scale. In this essay, I ask: could a planetary-oriented public sphere be thought out in line with what postcolonial theorist Gayatri Chakravorty Spivak refers to when she speaks about a radicalised dialogical practice of accountability — the exchange between publics, between inhabitants and migrants or between the North and the South, for example, as "a two-way road" following the "imperative to re-imagine the subject as planetary accident".[6] Within that worldly, and not simply nation-state framework, both the "dominant" and the "subordinate" supportively rethink themselves as "interpellated by planetary alterity, albeit articulating the task of thinking and doing from different 'cultural' angles", to address requirements that interpellate all of us, "as giver and taker, female and male, planetary human beings".[7]

With that question in mind (could a planetary-oriented public sphere be rethought as a radicalised dialogical practice of accountability?), I want to examine a work that explores the aesthetic challenge of dialoguing between worldviews: Igloolik Isuma Productions' video-and-webcasts intervention in the Canada Pavilion during the *58th Venice Biennale* in 2019 (Figure 1) — a work that bridges two planetary predicaments: forced migration and environmental degradation (namely, biodiversity loss as part of the ongoing sixth mass extinction and global warming, as well as any cause-and-effect relation among these two types of environmental degradation). To enable that dialogics, it proposes and enacts a renewed public sphere, one that partakes in, but significantly troubles, as I will show later, Jürgen Habermas's universalist, rationally-deployed and media-skeptic notion of the public sphere as a realm of social life where public opinion takes shape through deliberations between subjects who "come together as a public" around matters of common concern. Isuma's public sphere comprises a media encounter between an Inuit counterpublic and western publics (or publics from the Global North, south of the Arctic North, mainly south of Nunavut), following a storytelling relation between "oral delivery and aural reception" that privileges listening over simply voicing or speaking.[8] My discussion of the work aspires to demonstrate that its coarticulation of two aesthetic strategies of coexistence — planetary mediality and listening — constitutes a unique counterpublic sphere, whose modus operandi is to relate worldviews, rather than simply multiplying or dividing them as within the predominant paradigm of cyberbalkanization. What is key here is the nature of the speaker-listener encounter, as an encounter that is sustained by the Inuit Qaujimajatuqangit principles of mutual respect, commonality, collaboration, acquisition of knowledge, and care for humans and nonhumans (the land, animals, and the environment), but also by the disabling of what Stó:lō scholar Dylan Robinson compellingly calls "hungry listening" — a form of listening that systematically ends up reinforcing the colonial denial of Indigenous difference. Key here is the requirement to search for a type of listening that keeps the incommensurability of Inuit and non-Inuit cultures alive. The strength of Isuma's intervention lies precisely in its ability to foster planetary mediality *and* listening so that the migrant situation and one of its central preconditions or pitfalls (environmental degradation) be brought to the fore as a matter of discussion and collaborative struggle. My essay first describes the intervention, briefly contextualises it within the history of colonialism in Canada, and then examines its media and listening practices. Ultimately, it aims to rethink coexistence in light of Isuma's elaboration of a counterpublic sphere informed by Inuit Qaujimajatuqangit principles and the assumption of the incommensurability of cultures.

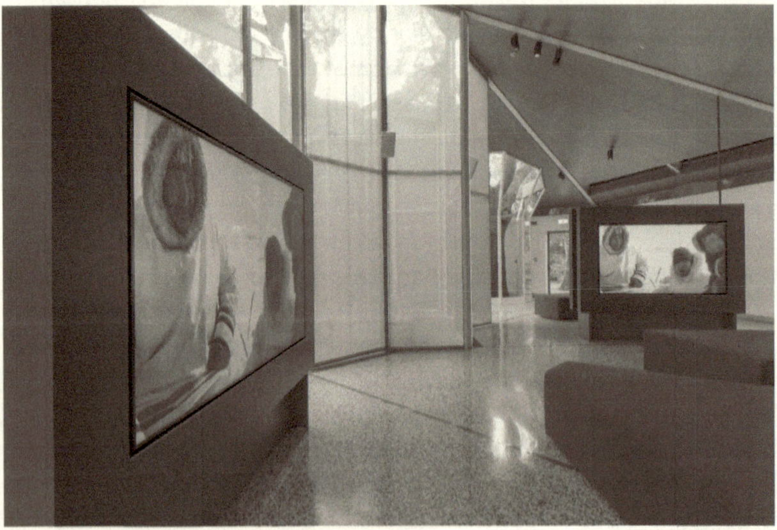

Figure 1. Venice installation photo: Isuma — *Isuma,* Canada Pavilion, *Venice Biennale 2019* (Photo: Francesco Barasciutti). Image courtesy of National Gallery of Canada and Isuma Distribution International.

2019 *Venice Biennale* and its publics

Isuma (Inuktitut Syllabics: ᐃᓱᒪ) is a collective of Inuit creators — the first predominantly Inuit collective invited to exhibit their work in the Canadian Pavilion in Venice. Co-founded in 1990 by Zacharias Kunuk, Paul Apak Angilirq, Pauloosie Qulitalik, and Norman Cohn, and primarily devoted to the production of independent video art, it has also helped establish several Inuit media institutions, including: an Igloolik-based Nunavut independent television network center (NITV); a collective of women filmmakers (Arnait Video Productions); Artciq (a youth performance-oriented collective); SILA (an e-learning website about Inuit culture); IsumaTV (a website for Indigenous media art); and Digital Indigenous Democracy (an internet network, whose main mission is to inform and consult with Inuit communities about the development of the Baffinland Iron Mines Corporation and other resource projects).[9] These media undertakings elaborate and institutionalise the digital extension of Inuit storytelling as a form of oral history transmitted by Elders to younger generations — a process increasingly understood as a means of empowerment whose effectiveness lies in the listening activity and multi-perspectivism it entails.[10] As best summarised by Inuk artist, filmmaker, and curator asinnajaq, Isuma's projects:

> *share stories of Inuit who have been disempowered, and who find power by being given the opportunity to speak up. There isn't one simple truth in the world, so understanding a subject requires listening to many perspectives. Isuma isn't the one and lonely voice for Inuit, but they do cover a vast array of topics, including relocations, climate change and international social politics. [...] In the process of producing historically accurate narrative films, the knowledge of Elders is being shared, heard and valued.*[11]

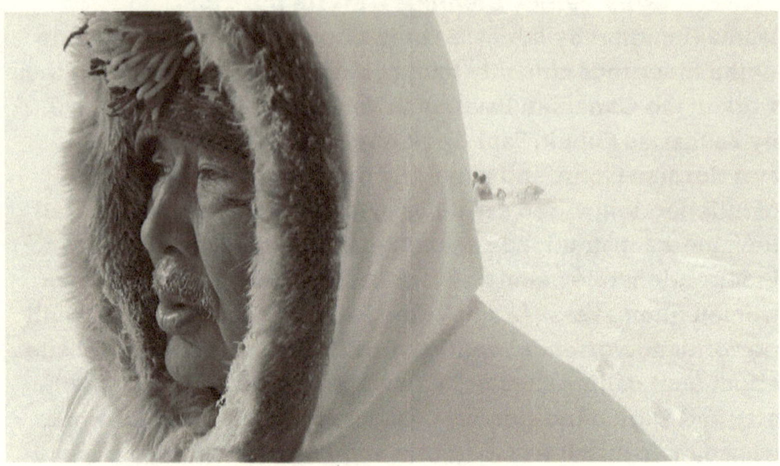

Figure 2. Film still: *One Day in the Life of Noah Piugattuk,* 2019. Image courtesy of Isuma Distribution International.

The Canadian Pavilion introduced two new works by Isuma Productions: a feature-length video in Inuktitut and English (with English and French subtitles), entitled *One Day in the Life of Noah Piugattuk* (ᓄᐊ ᐱᐅᒑᑦᑐᑉ ᐅᓪᓗᓂᓚᐅᖅᑕᖕᒪ, 2019) (Figure 2); and a series of four webcasts, titled *Silakut Live from the Floe Edge* (ᓯᓚᒃᑯᑦ ᓴᖅᑭᔮᖅᑐᑦ ᓯᓈᓂ, 2019). Both the video and the livecasts were screened in the pavilion, but could also be viewed online on IsumaTV, as well as in different galleries in Canada. The *Silakut* livecasts were held on 8, 9, 10, and 11 May. However, it is the joint presentation of the video and the webcasts that makes this intervention crucial, not only as an artistic response to the intertwinement of the migrant and environmental predicaments of the twenty-first century, but also as a substantial redefinition of the public sphere. Considered together, they affirm Inuit difference *and* connection (the encounter between members of the Inuit community, as well as between the North and the South) as a necessary combination for the struggle against environmental degradation. That upholding is a response to the growing precarity of the Igloolik community whose existence is threatened by biodiversity loss and global warming — environmental degradations identified in the webcasts as resulting from and caused by ice melt, as well as the development of the Mary River Project, an open-pit iron mine operated by the Baffinland Iron Mines Corporation in the Mary River area of Baffin Island, Nunavut. The company's plans for a phase-two expansion was scheduled to be heard by the Nunavut Impact Review Board (NIRB) in Iqaluit in the summer/fall of 2019, when Isuma's work was being shown and webcasted.[12]

Both the video and the livecasts situate the public sphere at the center of Inuit life. In the 112-minute long 4K digital video, *One Day in the Life of Noah Piugattuk,*[13] Inuk hunter Noah Piugattuk, surrounded by his band and a white man called the "boss" — an agent of the government, assigned to get Piugattuk and his band to move to a settlement housing development and send their children to school so that they could eventually get jobs and "make money" — meet at Piugattuk's hunting camp (Figure 3). The boss also invokes the context of the "war" (i.e. the Cold War) as justifying his request for delocalization-and-relocalisation. Staged in 1961 and shot on location in Kapuivik, north Baffin Island where Piugattuk and his band semi-nomadically lived and hunted, the docudrama is based on the life of Noah Piugattuk and the 1950–60s colonialist establishment of settlements through forced migration. Most of the video — and this is important when speaking about the constitution of publics — centers on the conversation, translated by an Inuk interpreter, between Piugattuk and the boss. They talk; they hear one another; they deliberate; they are publics to each other although in a two-way dialogue that is far from being dialogical, ruled as it is by the hierarchy of power securing the coloniser-colonised relation (Figure 4). Their statements are translated, yet often mistranslated or approximatively translated by the interpreter sitting between them. The deliberation ends when Piugattuk refuses to accept the boss's proposition. "I wanted to look at the moment that they [the Inuit] were told to move", says Kunuk. "They were saying, 'We don't want to go anywhere. We don't want to move'. But they were told they had to. So that's what we're looking at".[14] While Piugattuk said no to the move, his was a unique voice amidst the Inuit population whose destiny mainly took the form of imposed displacement.

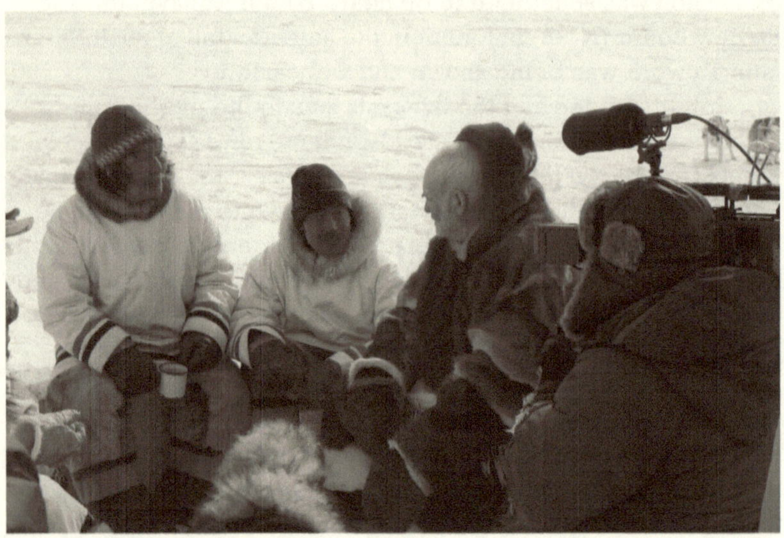

Figure 3. On-set photo: *One Day in the Life of Noah Piugattuk,* 2019 (Photo: Levi Uttak). Image courtesy of Isuma Distribution International.

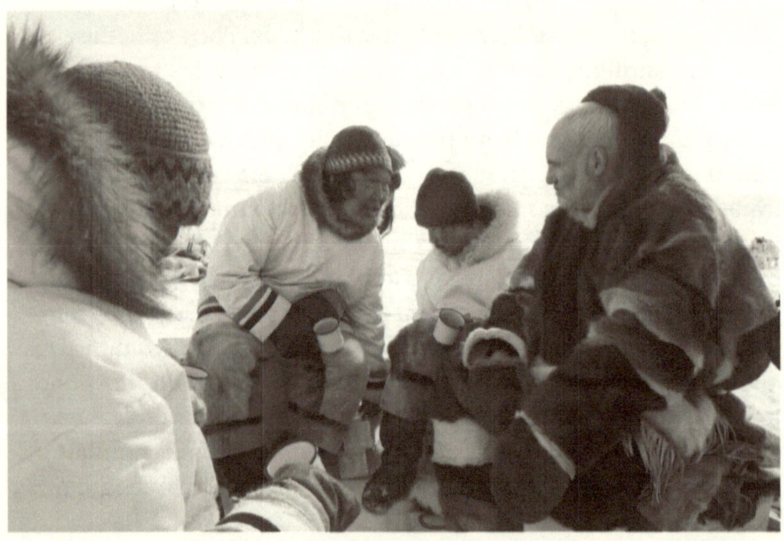

Figure 4. Film still: *One Day in the Life of Noah Piugattuk,* 2019. Image courtesy of Isuma Distribution International.

58 years later, filmed from within the Igloolik area, the four *Silakut Live from the Floe Edge* webcasts (Figure 5) capture and transmit another public sphere in the making: a counterpublic sphere. They show Kunuk sitting inside a cabin with Elders (as well as at least one member of the younger generation) from the Igloolik community. Gathered together, each member of the group talks one after the other, recalling memories of childhood, telling stories about human and shaman relationships, sharing their knowledge of different traditional cultural practices (including string games and drum dancing). Kunuk progressively invites them to talk about the development of the Mary River Project (more on this project below) and its impact on the community. The webcasts also present archives on past Inuit life; they transmit shots of the land, the floe edge where land meets the sea, as well as the film-crew and hunters active on the land, especially in the webcasts of 9 and 10 May when seal hunting is being filmed live — abandoned on the 9 because of the melting ice making it too thin to hunt, but resumed and successful on the 10. Describing *Silakut,* Kunuk insists on the imperative to webcast the community's environmental concerns about the Mary River Project as well as the melting of ice, implicitly echoing Inuit activist Sheila Watt-Cloutier's climate-change-informed call for "the right to be cold":[15] "Silakut means 'through the air.' [...] We plan to film live at our floe edge, from the ice and the sea, where hunters hunt seals, and broadcast halfway around the world to Venice. [...]

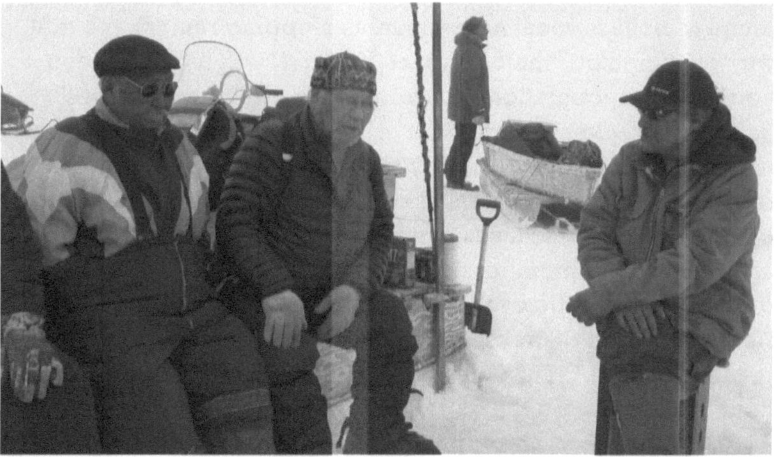

Figure 5. Screenshot of Silakut Live: *Silakut Live,* 2019. Image courtesy of Isuma Distribution International.

The land is melting, and we want to show that this summer".[16] The Mary River Property — I will be brief here — comprises a complex of nine-plus high-grade iron ore deposits (the proportion of iron contained in the ore at the site). The original land claim over the sediments was acquired by Baffinland Iron Mines Corporation in 1986 to develop a mine on the property. This became the Mary River Project, which expanded in 2012. Key to its ongoing development was the Inuit Impact and Benefits Agreement signed in 2013 with the Qikiqtani Inuit Association (QIA), which led to the federal approval of the project — an agreement that ensures benefits from the corporation's operation flow to nearby communities in North Baffin. The agreement allows Baffinland exploration and resource development rights to 170 km^2 of Inuit-owned land adjacent to the mine site. Baffinland Iron Mines is now in the process of seeking approval of its Phase 2 expansion to double its iron ore production. The plan is to double, and eventually triple production and export (from producing an estimated 4.2–6 million tons of iron ore to 12 million tons a year), and to construct a 110 km railway to carry the iron ore from the mine to Milne Inlet, near Pond Inlet, Nunavut, where it will ship the ore internationally. The project has raised significant environmental concerns within the scientific community, as well as within Inuit communities. These include concerns over the effect of freighters on the ice necessary for the survival of marine mammals (notably, the walrus and the narwhal — an arctic-dwelling whale that relies on sound to navigate, communicate, and find its prey but now found to be less vocal near the mine shipping routes); the railway's anticipated "major impact" on the North Baffin caribou herd "whose population is currently at a critically low level"; Baffinland's acknowledgement of fuel spills and water contamination; the company's inability to demonstrate its capacity to reduce greenhouse gas emissions — emissions considered to be one of the main causes of human-induced climate change; and claims from members of Inuit communities of a loud hum or buzz-soundscape evolving from within the Fury Strait and Hecla Strait and distressing the sea mammals the communities rely on for food.[17] The Nunavut Impact Review Board (NIRB) held its hearings on the expansion in Iqaluit in November 2019. The hearings were transmitted on IsumaTV by Digital Indigenous Democracy. Kunuk's plan was to film the proceedings of the NIRB meetings and to hold interviews with the intervenors.[18] The hearings were initiated but have been suspended. The conflict is still ongoing.

Key here is how media in the *Silakut* webcasts are used as a means to make the environmental costs of extractivism public — extractivism understood here as a mode of accumulation; the process of extracting large quantities of natural resources from the Earth, mainly for export.[19] At the beginning of each webcast, and after each pause, Kunuk explicitly welcomes the public (which is always necessarily a shifting Inuit and Qallunaat/non-Inuit public in and outside Venice, listening live, or, in its archived version, to the webcast from anywhere in the world) and invites them to listen. When asked about the Mary River Phase 2 Project, each member expresses — although some more forcibly than others — their distrust of the planned expansion. What is abundantly voiced is the project's depreciation of the interdependency between the humans and animals living in Baffinland; as well as the endangering of the animals (fish, walrus, narwhal, caribou) and what this represents for a community that has traditionally hunted for food. Examples of such statements include: "They don't know nothing about what they will be doing [...] our land, our animals [...]"; "Our land has much resources [...] People, we eat from land and water [...] the metal will go into our bodies. The things we eat will have to be tested"; "If they continue, the animals won't be there anymore. That's all I have to say"; "I already know what's going on [...] the high area is much colder than the lower area, and if they combine [...] it's a bad sign". Similar to the dialogic structure of *One Day in the Life of Noah Piugattuk,* a translator, now off-screen and addressing the off-screen audience exclusively, translates from Inuktitut to English, yet only approximatively — showing that communication between publics is never straightforward or transparent. And yet, both the video and the livecasts value listening as much as, or even more than speaking. Listening *enables* speaking, insofar as it provides the necessary silence for each individual to think and then express him- or herself. Thus, the dialogue is never direct and is not particularly conversational — the comments are answers to Kunuk's questions, but not a back-and-forth discussion between the members of the group onscreen: each member gives his or her perspective, following the oral-delivery-and-aural-reception tradition of storytelling. We, the audience, are positioned as listeners in the same way: as guests we are invited to hear the different worldviews articulated from within the Igloolik community, and to act accordingly, if so desired.

The reinvention of the counterpublic sphere, or, how to make public the historical link between forced migration and environmental degradation

The video and the webcasts were transmitted together in the same space (the Canada Pavilion) and period. They were thought out together and are, in fact, inseparable. That jointness is crucial insofar as it establishes a historical link between forced migration and environmental degradation — biodiversity loss and global warming, with a special reflection on extractivism as a central human activity responsible for that degradation. Inuit are not simply climate victims, but first and foremost victims of colonial displacement and dispossession. What is being exposed here are the ongoing consequences of the *grand renversement* ("great reversal") lived by the Inuit whose mode of existence was essentially semi-nomadic — a renversement that reached its peak in 1939 when the Inuit became a federal responsibility, and then in the 1950s and 60s when they were moved off the land to be relocated in permanent settlements. In these settlements, Inuit were subjected to assimilation policies imposing the "Canadian way of life": pressured to abandon their traditions, they became increasingly dependent on the government for education, health care, police force, housing, food, work, and other services. As subsistence hunters, they had lived interdependently; they depended both on nature and on each other to survive in the harsh Arctic climate. That interdependency decreased as their dependence on the Qallunaat increased.[20] Inuit existence was transformed, in many cases through forced migration, into a sedentarily lifestyle following a colonialist imposed acculturation logic of disconnection from land, culture, and community.[21] As the work of anthropologist Hugh Brody and historian Colin G. Calloway has shown, colonialism has had a leading effect on Inuit communities in its strive to eliminate their interdependency-based kinship system, hunting practices, and oral tradition (including cosmology, animism, shamanism, and storytelling), which had been pivotal to their survival in the Arctic north.[22] These transformations of human-nonhuman interdependency all relate to a depreciation of interrelatedness promoted in Inuit culture — values of, and beliefs in connectedness and belongingness that have been identified and now documented by Inuit Elders as intrinsic to Inuit Qaujimajatuqangit (IQ), an oral tradition transmission body of "beliefs, laws, principles, values, skills, knowledge and attitudes" passed from generation to generation, which serves today as an educational framework for curriculum in Nunavut educational institutions.[23] As stated by Inuk storyteller and language educator Mark Kalluak, "Inuit Qaujimajatuqangit means knowing the land, names, locations and their history. It also means knowledge of the Arctic environment — of snow, ice, water weather and the environment that we share. It encompasses being in harmony with people, land and living things — and respecting them".[24] That land-oriented body of knowledge outlines eight fundamental principles, including: mutual respect, collaboration and care for humans and nonhumans (the land, animals, and the environment). These principles are highly relevant to Igloolik's and, more largely, Nunavut's environmental protest against the Mary River Project: the protest's major demand is that Baffinland Iron Mines recognise the importance of IQ in its development of the mine.[25] Acculturation is, therefore, not a completed *renversement.* Inuit have voiced and are voicing their concerns about the environmental deterioration of their land — re-inscribing, as it were, Piugattuk's resilience. The webcasts are explored as media of resilience and resistance, as well as media of reclamation and transmission of disappearing traditions; they widen the public sphere by allowing publics to meet around Mary River's extractivist project and an alternative vision of the environment.

Notice, however, how the speakers in the webcasts never simply blame the South — they question the activities of the multinational company sustaining the Merry River Project, as well as the government, but the point of the webcasts is to speak about the environmental problem and make it as public as possible. It seeks a public sphere. It seeks to resume and reverse the 1960s sphere represented in *One Day in the Life of Noah Piugattuk.* Some members of the group mention — often as a statement — how the people from the South could help fund their cause; but it is never about the Inuit saying that environmental degradation is a condition lived in the same way by everyone on the planet; they mostly insist on this being an Inuit cause — they are the actors and not simply the victims seeking pity or empathy from the South. This dialogical approach is "consistent" with the aims of Indigenous self-determination.[26]

It is their cause, and their cause needs, strategically, to be heard by the largest public possible, including Inuit and non-Inuit peoples. Hence the value of the livecasts, which can potentially be heard from everywhere and by anyone on the planet, while being firmly sited in the floe edge in the Igloolik area, to reestablish the interdependency colonialism works to disconnect (Figure 6).[27] Implied, of course, is the hope that the public realises that this cause is not only worthwhile, especially in light of the responsibility of the South for its forced displacement of Inuit populations, but also beneficial to all. The voices from the South were not heard during the webcasts — the public was encouraged to attend and for now it is too early to measure the scope of their responses.

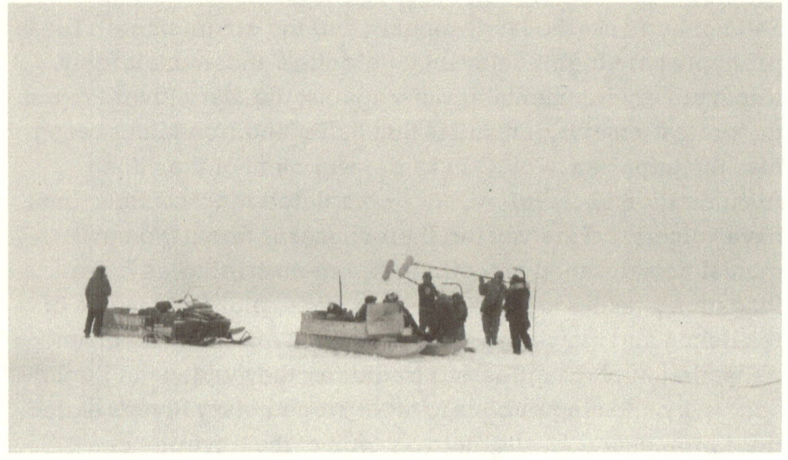

Figure 6. Screenshot of Silakut Live: *Silakut Live,* 2019. Image courtesy of Isuma Distribution International.

What is Isuma's 2019 video-and-webcasts intervention if not a public sphere instantiation — identified as such by Isuma's Norman Cohn,[28] although one that substantially redefines its Habermassian deployment? Let us recall Habermas's conceptualisation of that particular form of critical publicity. In *The Structural Transformation of the Public Sphere* (first published in 1962 and translated into English in 1989), Habermas defined the modern public sphere as a realm of social life where public opinion takes shape. This realm forms itself around rational deliberations between individuals who "come together as a public" as they debate on matters of general interest and common concern.[29] Its ideal type is the eighteenth-century bourgeois public sphere, whose efficiency lay in its capacity to act as a normative principle of democratic legitimacy, producing public opinion that influenced political action against the domination of the state. In subsequent revisions, Habermas emphasised the role of deliberative language and communicative rationality in the consolidation of the public sphere, which he redefined as "a network for communicating information", where participants rationally express their points of view by adopting positions and assuming illocutionary obligations that support mutual speech acts.[30] The Habermassian formulation of the public sphere has been contested from the start. Critics have questioned its presumed universalism, as well as its rationalist structure. Critical theorist Nancy Fraser has shown that the bourgeois public sphere was constituted through a considerable number of exclusions — women, in particular, and other social groups; the excluded eventually grouped together to form counterpublics and formulate oppositional claims based on their own identities and interests.[31] Philosophers Oskar Negt and Alexander Kluge have disclosed the interdependency between the bourgeois public sphere and the proletarian counterpublic sphere.[32] Political philosopher Chantal Mouffe has contested Habermas's rationalistic model of argumentation, to propose instead an agonistic model where antagonism is *the* necessary passion of politics.[33] Media scholars have shown that the interpersonal relationships composing the public sphere were much more mediated than Habermas initially presumed, and that the development of mass media does not necessarily lead to the decline of the public sphere.

How to listen?

In light of these critiques, what remains of the public sphere today, and what is to be saved from it? How can it be reinvented to address forced migration, its colonial unfolding, as well as its environmental causes and consequences? How can it be rethought as a coexistence based on Inuit Qaujimajatuqangit principles? Much more multiple, porous, passionate, mediated, local, and mutable than initially formulated, certainly much weaker in a post-factual society, some key components of this critically reformulated public sphere resurface in Isuma's *Silakut Live from the Floe Edge* webcasts, especially when considered in relation to the *One Day in the Life of Noah Piugattuk* video.

Mediality — more specifically, the use of webcast technology — is brought to the fore as a vital means of creating an enlarged counterpublic sphere: by interpellating both the North and the South (anyone from the planet, in fact) without losing the historical, geographic, economic, social, and cultural specificities of Inuit perspectives, it problematises cyberbalkanization. That media-enabled interdependency of publics (a pivotal value of IQ) brings us to the second aesthetic strategy constitutive of Isuma's counterpublic sphere: listening. The video and webcasts define the public sphere as a forum within which to speak up, but also, and more fundamentally, as an aural reception. Some webcasted statements from the Igloolik Inuit community members are rather explicit in this regard: "Do they hear what the Inuit want?" In their individual statements, they advocate for a connection between publics not only across difference but through listening: "We would have to have meetings ourselves [...] our people [...] we would have to expect something from us for ourselves and not people who want to land"; "People would have to start helping each other more and negotiate with each other for all that to stop"; "Since they are looking for iron or in this area we are trying to present this from happening and we will chat about it and you guys just listen"; "We are live right now all over. We are showing this in Venice. We want these people to know and we want you guys to watch when we are live [...] We want to get help from the people instead of just saying this"; "I want people to know"; "Since they are looking for iron or in this area we are trying to prevent this from happening and we will talk about it and you guys just listen".

Listening might well be the forgotten practice of our times. Could it be envisaged as a way to weaken post-factuality — a mode of listening to the other's story which holds open, as suggested by philosopher Jean-Luc Nancy, the threshold between sending and resending, sense and signification? [35] The work of First Nations scholar Dylan Robinson, who specialises in the study of music aesthetics and Indigenous artistic and cultural practices, shows how listening cannot be valued as a form of mutual respect, connectedness, and reciprocity in and of itself. Listening encounters, between the speaker and the listener, the musician and his or her public, are mobilised by positionalities — by "how we listen as Indigenous, settler, and variously positioned subjects".[36] In his recent book, *Hungry*

Listening: Resonant Theory for Indigenous Sound Studies (2020), Robinson asks two questions that are fundamental to any instauration of a public sphere: how do we listen?; and how to listen? The settler's listening is mainly a "hungry listening", one that persists even in inclusive and collaborative performing practices between Indigenous and non-Indigenous musicians, as well as between Indigenous musicians and settler audiences. It is a form of listening that reproduces colonial, colonialist, and neo-colonial listening that systematically seek to "civilise" and assimilate Indigenous voices — a mode of perception "that has been imposed on Indigenous people who grew up in residential school, boarding school, and day school systems; [...] who have been disfranchised by [...] forced migration".[37] Following journalist and environmental activist Naomi Klein's insight into extractivism — as being not only a mining and drilling practice, but more importantly a colonial assimilative "mindset", in which not only land but Indigenous peoples and knowledge are seen as resource "to be mined" — Robinson defines hungry listening as an extractivist practice.[38] Settlers (a term generically used to refer to the persistence of colonialism in the West) absorb what is "digestible" to them in Indigenous music: sonorities, song content, and stories that we expect (from trauma to healing to reconciliation) and "fit" our sensibility without displacing our positionality.[39] Seeking reciprocal performances where listening encounters between Indigenous and settler positionalities resist extractivism, Robinson proposes that musicians and listeners "attend to being between [...] ontologies and sound worlds" while hearing and sensing the land through the song, so that the incommensurability of Indigenous and settler cultures be preserved.[40]

Robinson's critical questioning of hungry listening asks that we complicate our understanding of the intersubjective relations sustaining the North-South public sphere elaborated by Isuma. Could it be said that the webcasts' off-screen publics (a public whose composition changes with each new viewing) are invited to temper their perceptual avidity? Without that repositioning, Igloolik is destined to become a mere sub-public (a sub-public isolated from other sub-publics that fail to listen to each other), or a public that only manages to confirm or infirm what the settler expects of it, such as the "boss" facing Piugattuk. Isuma's double intervention seems to me to be more sophisticated than these two orientations, insofar as it never abandons the question "how to listen?" Three aesthetic strategies (perhaps more) work to trouble hungry listening. Firstly, the decentering of the *Venice Biennale*'s public, enabled by the use of web technology, has managed to bring not only Venice but potentially anyone (any settler) on the planet to Igloolik. Secondly, the storytelling structure of the webcasted dialogues insists on the necessity to listen to the other; it proposes the silence of listening as a pause that encourages a more attentive form of hearing before any materialisation of speech. And thirdly, the interpreters' mistranslations make manifest the historical failure of communication between Inuit and non-Inuit communities, but it also slows down the assimilationist intensity characteristic of hungry listening. These strategies have the potential — lucidly described by Robinson — to preserve the incommensurability of Indigenous and non-Indigenous cultures as they come together to constitute an enlarged dialogical public sphere. Especially in the context of cyberbalkanization and post-factuality, in which it has become so easy to deny the major (migration, environmental, or other) crises of the twenty-first century, it is imperative to account for artistic practices — Isuma's in particular — that invite us to be listening publics from (let us follow Spivak here) *"planetary discontinuity".*[41] The enlarged, potentially planetary public sphere, reimagined by Isuma, comprises publics from the North and the South; these publics are interpellated by an alterity that ceases to be an Indigenous fatality. Each public is encouraged to think about and act on — from different cultural viewpoints — the primordial crises they have in common.[42]

Notes

1. On the eradication of the other in the context of the migrant crisis, see Daniel Trilling, "Five Myths About the Refugee Crisis", *The Guardian,* 5 June 5 2018, https://www.theguardian.com/news/2018/jun/05/five-myths-about-the-refugee-crisis; Étienne Balibar, "Pour un droit international de l'hospitalité", *Le Monde,* 17 August 2018, 23, https://www.lemonde.fr/idees/article/2018/08/16/etienne-balibar-pour-un-droit-international-de-l-hospitalite_5342881_3232.html; and Achille Mbembe, Necropolitics (Durham, North Carolina: Duke University Press, 2019).

2. Michel Agier, *Définir les réfugiés* (Paris: Presses Universitaires de France, 2018), 5.

3. Nicholas De Genova, "The Borders of 'Europe' and the European Question", in *The Borders of "Europe": Autonomy of Migration, Tactics of Bordering,* ed. Nicholas De Genova (Durham and London: Duke University Press, 2017), 9.

4. Craig Calhoun, "The Idea of Emergency: Humanitarian Action and Global (Dis)Order", in *Contemporary States of Emergency: The Politics of Military and Humanitarian Interventions,* eds. Didier Fassin and Mariella Pandolfi (Brooklyn, New York: Zone Books, 2010), 30; and Michel Agier, *Managing the Undesirables: Refugee Camps and Humanitarian Government* (Cambridge, UK, and Malden, MA: Polity Press, 2011).

5. On cyberbalkanization, see especially one of the first articles written on that phenomenon: Marshall Van Alstyne and Erik Brynjolfsson, "Could the Internet Balkanize Science?" *Science* 274, no. 5292 (November 1996): 1479–1480.

6. Gayatri Chakravorty Spivak, "Imperative to Re-imagine the Planet", in *An Aesthetic Education in the Era of Globalization* (Cambridge, MA: Harvard University Press, 2009), 339 and 348.

7. Spivak, "Imperative to Re-imagine the Planet", 347 and 350.

8. On this specific storytelling relation, see Jo-Anne Archibald, *Indigenous Storywork: Educating the Heart, Mind, Body and Spirit* (Vancouver, BC: The University of British Columbia Press, 2008), 7.

9. "Silakut: Live from the Floe Edge", *Art Gallery of Alberta,* accessed 19 June 2019, https://www.youraga.ca/whats-happening/calendar/silakut-live-floe-edge; and "Making Independent Inuit Video for 30 years", *Isuma,* accessed 9 December 2019, http://www.isuma.tv/isuma.

10. Katarina Soukup, "Report: Travelling Through Layers: Inuit Artists Appropriate New Technologies", *Canadian Journal of Communication* 31, no. 1 (2006), https://www.cjc-online.ca/index.php/journal/article/view/1769/1889.

11. asinnajaq, "Isuma Is a Cumulative Effort", *Canadian Art* (Spring 2019), https://canadianart.ca/features/isuma-is-a-cumulative-effort.

12. Leah Sandals, "Zacharias Kunuk Speaks on Isuma's Venice Biennale Project", *Canadian Art* (8 May 2019), https://canadianart.ca/news/zacharias-kunuk-speaks-on-isumas-venice-biennale-project.

13. *In One Day in the Life of Noah Piugattuk* (2019), Noah Piugattuk is played by actor Apayata Kotierk; Isumataq (the Boss) is played by Kim Bodnia.

14. Sandals, "Zacharias Kunuk Speaks on Isuma's Venice Biennale Project".

15. Sheila Watt-Cloutier, *The Right to Be Cold: One Woman's Story of Protecting Her Culture, the Arctic and the Whole Planet* (Toronto: Penguin Random House Canada, 2015).

16. Sandals, "Zacharias Kunuk Speaks on Isuma's Venice Biennale Project".

17. "WWF Canada's Final Written Submission for the Nunavut Impact Review Board's Reconsideration of the Mary River Project Certificate for Baffinland's Mary River 'Phase 2' Proposal", WWF-Canada, accessed 4 December 2019, https://wwf.ca/wp-content/uploads/2020/03/wwf_final_written_submission_for_mary_river_phase_2_September-2019.pdf; "Mary River Mine", Baffinland, accessed 9 December 2019, https://www.baffinland.com/mary-river-mine/mary-river-mine; "Mary River Project", Qikiqtani Inuit Association, accessed 9 December 2019, https://www.qia.ca/about-us/departments/major-projects/what-is-the-mary-river-project; Beth Brown, "Baffinland Cuts Contracts, Leaves 96 Inuit Without Work", CBC News, 15 November 15 2019, https://www.cbc.ca/news/canada/north/baffinland-contracts-cut-mary-river-inuit-jobs-1.5361604; Meagan Deuling, "Baffinland Must Clarify Effects on Narwhal Before

Expansion of Nunavut Iron Ore Mine", *CBC News,* 26 April 2019, https://www.cbc.ca/news/canada/north/second-technical-meeting-for-baffinland-1.5111345; and Sara Frizzell, "Environmental Group Asks to Suspend Baffinland Mine's Approval Process", *CBC News,* 2 November 2019, https://www.cbc.ca/news/canada/north/baffinland-phase-2-hearings-oceans-north-1.5345336.

18. Sandals, "Zacharias Kunuk Speaks on Isuma's Venice Biennale Project".

19. Alberto Acosta, "Extractivism and Neoextractivism: Two Sides of the Same Curse", in *Beyond Development: Alternative Visions from Latin America,* eds. Miriam Lang and Dunia Mokrani (Quito: Transnational Institute, Fundación Rosa Luxemburg, 2013), 127.

20. Willem Rasing, *Too Many People: Contact, Disorder, Change in an Inuit Society,* 1822–2015 (Iqaluit, NU: Nunavut Artic College Media, 2017), 244.

21. Rasing, *Too Many People,* 224.

22. Hugh Brody, *The Other Side of Eden: Hunters, Farmers and the Shaping of the World* (New York: North Point Press, 2000); and Colin G. Calloway, *One Vast Winter Count: The Native American West Before Lewis and Clark* (Lincoln and London: University of Nebraska Press, 2003).

23. The Government of Nunavut, *Inuit Qaujimajatuqangit Education Framework* (Iqaluit, Nunavut: Nunavut Department of Education, 2007), 32–34, https://www.gov.nu.ca/sites/default/files/files/Inuit%20Qaujimajatuqangit%20ENG.pdf; and Joe Karetak, Frank Tester, and Shirley Tagalik, eds., *Inuit Qaujimajatuqangit: What Inuit Have Always Known to Be True* (Black Point, NS, and Winnipeg, MB: Fernwood Publishing, 2017), 225–230.

24. Mark Kalluak, "About Inuit Qaujimajatuqangit", in *Inuit Qaujimajatuqangit: What Inuit Have Always Known to Be True,* 41.

25. Email exchange with Norman Cohn, 16 January 2020.

26. Sandals, "Zacharias Kunuk Speaks on Isuma's Venice Biennale Project".

27. Arctic Kingdom, "What is the Floe Edge" (1 February 2019), https://arctickingdom.com/what-is-the-floe-edge.

28. Phone interview with Norman Cohn, 15 January 2020.

29. Jürgen Habermas, *The Structural Transformation of the Public Sphere: An Inquiry into a Category of Bourgeois Society,* trans. Thomas Burger and Frederick Lawrence (Cambridge, MA: MIT Press, 1991), 27.

30. Jürgen Habermas, *Between Fact and Norms: Contributions to a Discourse Theory of Law and Democracy,* trans. William Rehg (Cambridge, MA: MIT Press, 1996), 361.

31. Nancy Fraser, "Rethinking the Public Sphere: A Contribution to the Critique of Actually Existing Democracy", in *Habermas and the Public Sphere,* ed. Craig Calhoun (Cambridge, MA: MIT Press, 1992), 109–142.

32. Oskar Negt and Alexander Kluge, *Public Sphere and Experience: Toward an Analysis of the Bourgeois and Proletarian Public Sphere,* trans. Peter Labanyi, Jamie Owen Daniel, and Assenka Oksiloff (Minneapolis, MN: University of Minnesota Press, 1993).

33. Chantal Mouffe, *Agonistics: Thinking the World Politically* (London: Verso, 2013).

34. John B. Thompson, *The Media and Modernity: A Social Theory of the Media* (Stanford: Stanford University Press, 1995); and Manuel Castells, "The New Public Sphere: Global Civil Society, Communication Networks, and Global Governance", *The Annals of the American Academy of Political and Social Science* 616, no. 1 (March 2008): 78–93.

35. Jean-Luc Nancy, *Listening,* trans. Charlotte Mandell (New York: Fordham University Press, 2007).

36. Dylan Robinson, *Hungry Listening: Resonant Theory for Indigenous Sound Studies* (Minneapolis and London: University of Minnesota Press, 2020), 2.

37. Robinson, *Hungry Listening,* 3.

38. Robinson, *Hungry Listening,* 14.

39. Robinson, *Hungry Listening,* 50–51.

40. Robinson, *Hungry Listening,* 53.

41. Spivak, "Imperative to Re-imagine the Planet", 342.

42. Spivak, "Imperative to Re-imagine the Planet", 347 and 350.

References

Acosta, Alberto. "Extractivism and Neoextractivism: Two Sides of the Same Curse". In *Beyond Development: Alternative Visions from Latin America,* edited by Miriam Lang and Dunia Mokrani, 61–86. Quito: Transnational Institute, Fundación Rosa Luxemburg, 2013.

Agier, Michel. *Définir les réfugiés.* Paris: Presses Universitaires de France, 2018.

Agier, Michel. *Managing the Undesirables: Refugee Camps and Humanitarian Government.* Cambridge, UK, and Malden, MA: Polity Press, 2011.

Archibald, Jo-Anne. *Indigenous Storywork: Educating the Heart, Mind, Body and Spirit.* Vancouver, BC: The University of British Columbia Press, 2008.

Art Gallery of Alberta. "Silakut: Live from the Floe Edge". Accessed 19 June 2019. https://www.youraga.ca/whats-happening/calendar/silakut-live-floe-edge.

Arctic Kingdom. "What is the Floe Edge". Accessed 30 August 2020. https://arctickingdom.com/what-is-the-floe-edge.

asinnajaq. "Isuma Is a Cumulative Effort". *Canadian Art* (Spring 2019). https://canadianart.ca/features/isuma-is-a-cumulative-effort.

Baffinland. "Mary River Mine". Accessed 9 December 2019. https://www.baffinland.com/operation/mary-river-mine/mary-river-mine.

Balibar, Étienne. "Pour un droit international de l'hospitalité". *Le Monde,* 17 August 2018. https://www.lemonde.fr/idees/article/2018/08/16/etienne-balibar-pour-un-droit-international-de-l-hospitalite_5342881_3232.html.

Brody, Hugh. *The Other Side of Eden: Hunters, Farmers and the Shaping of the World.* New York: North Point Press, 2000.

Brown, Beth. "Baffinland Cuts Contracts, Leaves 96 Inuit Without Work". *CBC News,* 15 November 2019. https://www.cbc.ca/news/canada/north/baffinland-contracts-cut-mary-river-inuit-jobs-1.5361604.

Calhoun, Craig. "The Idea of Emergency: Humanitarian Action and Global (Dis)Order". In *Contemporary States of Emergency: The Politics of Military and Humanitarian Interventions,* edited by Didier Fassin and Mariella Pandolfi, 29–58. Brooklyn, NY: Zone Books, 2010.

Calloway, Colin G. *One Vast Winter Count: The Native American West Before Lewis and Clark.* Lincoln and London: University of Nebraska Press, 2003.

Castells, Manuel. "The New Public Sphere: Global Civil Society, Communication Networks, and Global Governance". *The Annals of the American Academy of Political and Social Science* 616, no. 1 (March 2008): 78–93.

De Genova, Nicholas. "The Borders of 'Europe' and the European Question". In The Borders of "Europe": *Autonomy of Migration, Tactics of Bordering,* edited by Nicholas De Genova, 9–15. Durham and London: Duke University Press, 2017.

Deuling, Meagan. "Baffinland Must Clarify Effects on Narwhal Before Expansion of Nunavut Iron Ore Mine". *CBC News,* 26 April 2019. https://www.cbc.ca/news/canada/north/second-technical-meeting-for-baffinland-1.5111345.

Fraser, Nancy. "Rethinking the Public Sphere: A Contribution to the Critique of Actually Existing Democracy". In *Habermas and the Public Sphere,* edited by Craig Calhoun, 109–142. Cambridge, MA: MIT Press, 1992.

Frizzell, Sara. "Environmental Group Asks to Suspend Baffinland Mine's Approval Process". *CBC News,* 2 November 2019. https://www.cbc.ca/news/canada/north/baffinland-phase-2-hearings-oceans-north-1.5345336.

Habermas, Jürgen. *The Structural Transformation of the Public Sphere: An Inquiry into a Category of Bourgeois Society.* Translated by Thomas Burger and Frederick Lawrence. Cambridge, MA: MIT Press, 1991.

Habermas, Jürgen. *Between Fact and Norms: Contributions to a Discourse Theory of Law and Democracy.* Translated by William Rehg. Cambridge, MA: MIT Press, 1996.

Isuma. "Making Independent Inuit Video for 30 years". Accessed 9 December 2019. http://www.isuma.tv/isuma.

Kalluak, Mark. "About Inuit Qaujimajatuqangit. In *Inuit Qaujimajatuqangit: What Inuit Have Always Known to Be True,* edited by Joe Karetak, Frank Tester, and Shirley Tagalik, 41–60. Black Point, NS, and Winnipeg, MB: Fernwood Publishing, 2017.

Mbembe, Achille. *Necropolitics.* Durham, NC: Duke University Press, 2019.

Mouffe, Chantal. *Agonistics: Thinking the World Politically.* London:Verso, 2013.

Nancy, Jean-Luc. *Listening.* Translated by Charlotte Mandell. New York: Fordham University Press, 2007.

Negt, Oskar, and Alexander Kluge. *Public Sphere and Experience: Toward an Analysis of the Bourgeois and Proletarian Public Sphere.* Translated by Peter Labanyi, Jamie Owen Daniel, and Assenka Oksiloff. Minneapolis, MN: University of Minnesota Press, 1993.

Qikiqtani Inuit Association. "Mary River Project". Accessed 9 December 2019. https://www.qia.ca/about-us/departments/major-projects/what-is-the-mary-river-project.

Rasing, Willem. *Too Many People: Contact, Disorder, Change in an Inuit Society, 1822–2015.* Iqaluit, NU: Nunavut Artic College Media, 2017.

Robinson, Dylan. *Hungry Listening: Resonant Theory for Indigenous Sound Studies.* Minneapolis and London: University of Minnesota Press, 2020.

Sandals, Leah. "Zacharias Kunuk Speaks on Isuma's Venice Biennale Project". *Canadian Art* (May 8, 2019). https://canadianart.ca/news/zacharias-kunuk-speaks-on-isumas-venice-biennale-project.

Soukup, Katarina. "Report: Travelling Through Layers: Inuit Artists Appropriate New Technologies". *Canadian Journal of Communication* 31, no. 1 (2006). https://www.cjc-online.ca/index.php/journal/article/view/1769/1889.

Spivak, Gayatri Chakravorty. *An Aesthetic Education in the Era of Globalization.* Cambridge, MA: Harvard University Press, 2009.

Thompson, John B. *The Media and Modernity: A Social Theory of the Media.* Stanford: Stanford University Press, 1995.

Trilling, Daniel. "Five Myths About the Refugee Crisis". *The Guardian,* 5 June 2018. https://www.theguardian.com/news/2018/jun/05/five-myths-about-the-refugee-crisis.

Van Alstyne, Marshall, and Erik Brynjolfsson. "Could the Internet Balkanize Science?" *Science* 274, no. 5292 (November 1996): 1479–1480.

Watt-Cloutier, Sheila. *The Right to Be Cold: One Woman's Story of Protecting Her Culture, the Arctic and the Whole Planet.* Toronto: Penguin Random House Canada, 2015.

WWF-Canada. "WWF Canada's Final Written Submission for the Nunavut Impact Review Board's Reconsideration of the Mary River Project Certificate for Baffinland's Mary River 'Phase 2' Proposal". Accessed 4 December 2019. https://wwf.ca/wp-content/uploads/2020/03/wwf_final_written_submission_for_mary_river_phase_2_September-2019.pdf.

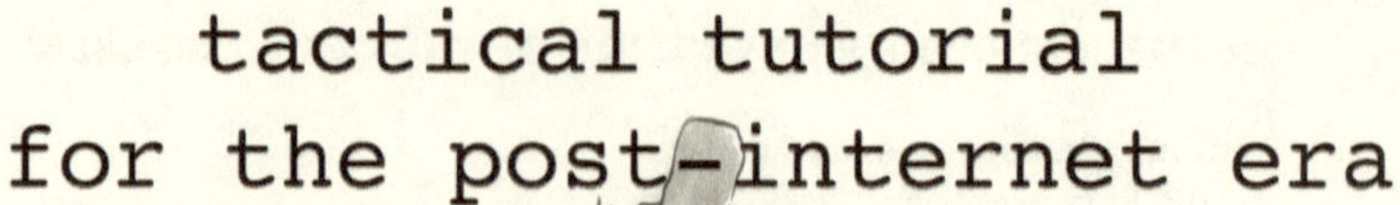

tactical tutorial for the post-internet era

Gregory Sholette

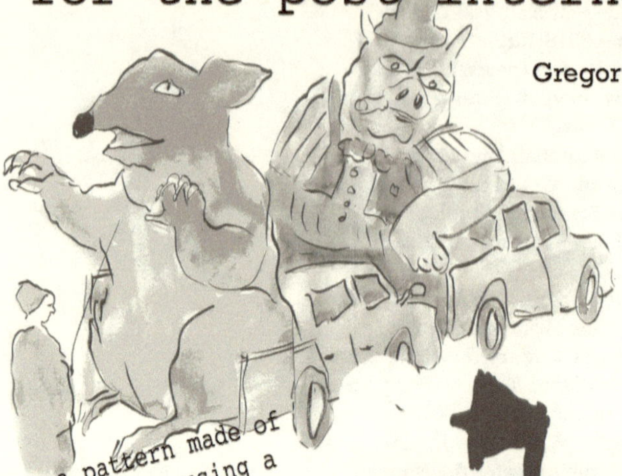

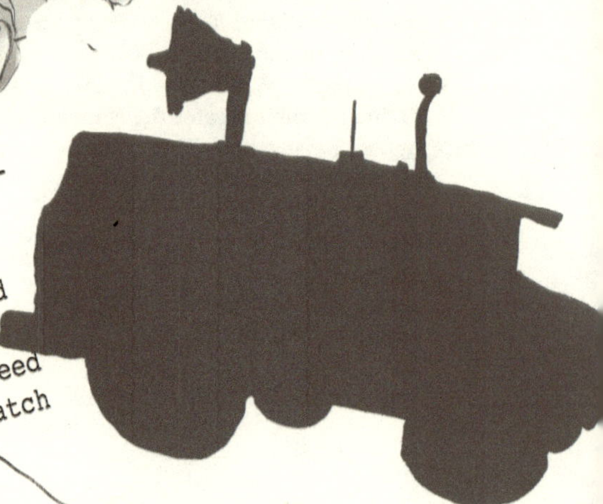

Patch together a pattern made of plastic shower curtains using a roll of silver duct tape. Add acrylic paint and attach a portable blower for an agit-prop inflatable. Facial makeup defies recognition software. Organized museum interventions press for rights. DIY street stencils need a lookout who knows how to watch for surveillance drones.

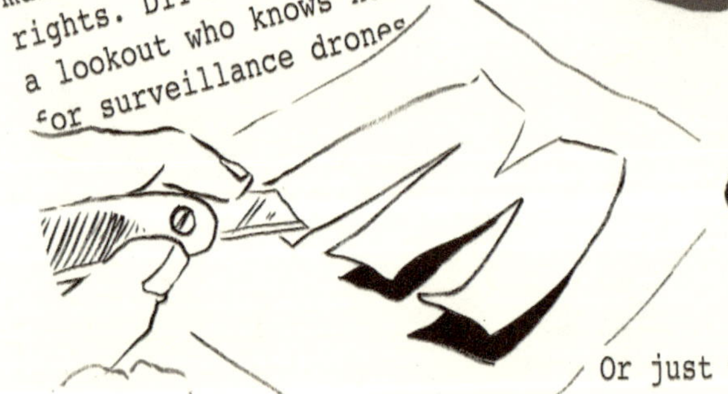

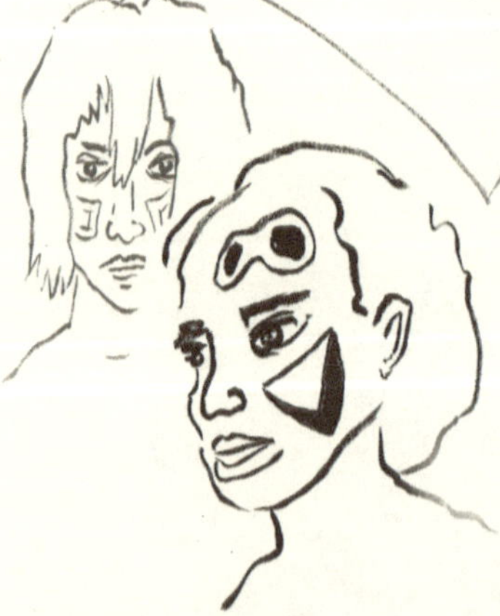

Or just take down their damn flag!

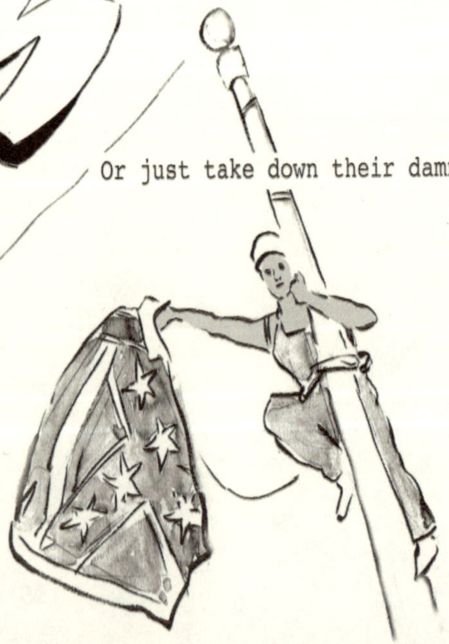

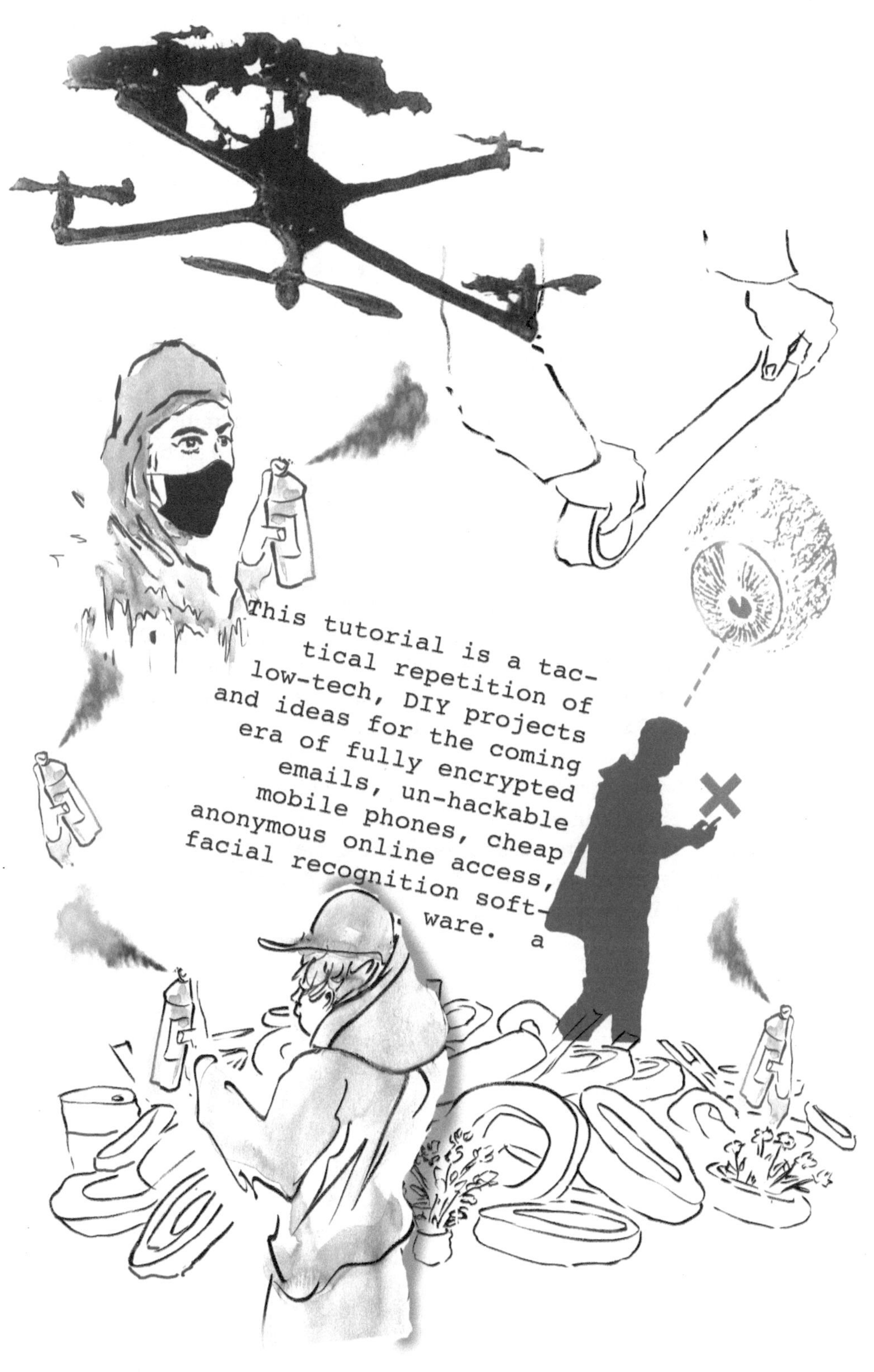

School of dissident studies

samiZine, *n.* (portmanteau)

A printed document or pamphlet that merges the informality of a fanzine with the hand-made, clandestine samizdat publication of the former mid-twentieth century Eastern Europe. A clear-cut example of a samiZine is the two-dozen underground newsprint tutorials produced by The School of Dissident Studies (SoDS) between 2024 and 2031. According to *PLASTIQUE* journal (Vol.4, No.1, 2061), The School reintroduced assorted DIY pre-digital tactics of resistance to the neo avant-garde diaspora in the aftermath of the Double City EMP event of 2022. As endowed Greenwald-MacPhee Professor Agata Craftlove points out, SoDS's samiZines singularly revitalised the all but lost knowledge of simple techniques for creatively disrupting everyday oppression, including the fabrication of anonymous inflatable sculptures, the cutting of street stencils, the application of anti-surveillance facial makeup, and other modes of untraceable direct intervention. The School's "Short the Future" campaign is also credited with partially liberating post-public spaces held captive by electronically insulated state and paramilitary xenophobe militias. See also ***Bad Deeds*** and ***Repulsive Aesthetics.*** **GLORYPEDIA,** first edition.

As part of an ongoing lexicon of imaginary idioms, my contribution to Fabricating Publics is a newly amalgamated artifact from an arcane future inventoried in the *Socially Engaged Art Glorypedia,* compiled by Gregory Sholette, Agata Craftlove, Karl Lorac, and TJ (www.themm.us).

Post-Truth as Bullying

Emily Rosamond

What happens to institutional critique in a moment of flat-out institutional attack?

How is it possible to critique institutions in this moment — is it a "post-truth moment", a "pandemic moment", a "crisis moment"? — without feeling like these days, institutions are really quite easy targets? Anyone can see: they're crumbling already. Institutions are under attack; institutions are sites of attack; institutions attract myriad modes of erosion. Budgetary crises force "difficult decisions" across art institution boardrooms. Changes of management seem like hostile takeovers.[1] High-profile political SNAFUs[2] reveal contempt for parliamentary process and established institutional procedures.[3] A pandemic pops along, like a litmus test revealing gaps in social welfare decades in the making. Everywhere, the feeling of the ship going down, of a system that doesn't work, of being on the cusp of an infrastructural breakdown. Or, maybe it's better to say being *in* such a breakdown — one unfurling, for the most part, infinitesimally slowly, like the shifting of continents — even if punctuated by the occasional (electoral) landslide.

Wide-ranging distrust of institutions persists; but much is transpiring, too, that's far worse than the institutions withering before our eyes. When the walls are caving in, how do you question "institutional authority" in the abstract — and for what? What winds are we witnessing anyway, ripping through "the institution": its boardrooms, its committee meeting cycles, its backwater filing systems, its decaying paperwork?

These days, London feels like a front-row seat for the macabre spectacle of institutional failure. What winds rip through institutions at the "margins" of the state — where bureaucratic fuck-ups, oversights, and wilful ignorance — perhaps best typified by endlessly dysfunctional, outsourced immigration proceedings, as with the Windrush scandal — place marginalised citizens in precarious relation to paperwork?[4] The endless malfunction of immigration procedures exacts a micro-political attack on subjects of the "hostile environment" — a bringing-up-the-drawbridge imaginary, carried out one lost bit of paperwork at a time.[5]

Meanwhile, what winds rip through the state's "centre" — where plutocrat-backed, would-be demagogues descend on Westminster, declaring an end to pointless institutional procedures? Brexit as institutional sabotage; Brexit as hostile takeover; Brexit as shorting the Pound. Maybe Brexit-as-sabotage speaks of a painful shift in class allegiances in the UK's Tory party: from the "regular rich" (business owners and the like, likely to be hurt by disruptions of their legal and bureaucratic continuities), to the super-rich — who, disaster capitalism-style,[6] presume to have little to lose and much to gain from widespread chaos and disruption, harnessed with a hedge fund manager's strategic foresight.[7] Parliaments legislate and prosecute to maintain some shred of adherence to procedure in face of this newly foregrounded, financialised disruption-logic. In the meantime, demagogue-ish, far-right politicians try to whip up factions of furious possible voters with social media-fuelled psyops. Parliamentary process hasn't caught up with this level of disruption-by-rote — a fact to which parliament's own 2020 Intelligence and Security Committee Russia report abundantly attests.[8] So, cast it off at all costs (the battle cry goes): this slow, cumbersome machine, delaying decisionist sensibilities, according to which a referendum outcome, or any other favoured directive, ought to be carried out quickly, as if by rote.[9]

What happens to critique (or criticality, for that matter)[10] in this moment of widespread attack? What kind of "object" could orient critique effectively, amidst an array of covert tactical actors (billionaire hedge fund managers, PR specialists, campaign strategists, and shareholders), endless puppeteering and pulling strings – without, on the other hand, oversimplifying the scene merely for the sake of concretising an object for critique to focus on? I would like to propose that the figure of the bully might be just such a provisional object. As a coercive sensibility corroding both institutional procedure and factuality generalises — to put it quite bluntly — bullying becomes the *modus operandi* of "post-truth". Thus, critical investigations of the figure of the bully could well play a foregrounded role in reinvigorating institutional critique and its concomitant practices.

The bully lodged in the institution, strong-arming people and calling the shots, becomes a "conceptual persona" of post-truth — a figure whose presence enunciates the weaknesses of institutional infrastructure and procedure.[11] The bully acts as supplement *and* sandpaper to that set of procedures — propping them up or eroding them as needed. The bully is an "anti-charismatic forcefield" enabling institutional attack. The bully is a foregrounded figure, peppered through tabloids and telly, widely circulating as an *image of institutional dysfunctionality.* Yet, in spite of its caricatured, feature-film forms, the bully isn't usually clear-cut around the edges. Often, it *fails to appear* as a figure separated from the ground of "business as usual". In a moment of widespread epistemic vice, the bully *figurates* (in other words, expresses and encapsulates an aspect of the zeitgeist as a figure) the mood of institutional attack that permeates the bureaucratic landscape, without necessarily being traceable to a decipherable point of origin.

Post-truth as coercion

Coercion eclipses factuality. The most insistent discourse wins. Entangled with any civilization's "truth procedures" is the possibility that the designation "factuality" carries an uncomfortable relationship to manipulation and coercion.[12] To come to be composed and consecrated as fact, in many instances, presupposes the active suppression of contradictory orderings of information and ideas that might threaten a hegemonic worldview. The term post-truth may well be limited (even if provisionally useful), insofar as it seems to fetishise "post-ness" — implying that the current condition is entirely new, as if people haven't had to weather massive disinformation campaigns before; or given its proclivity to incite wistful thinking about some erstwhile, "more factual" past. The phrase "post-truth as bullying" seems less to me like a stable, lasting thought, and more like an urgent, decaying proposition with a sharp sting and a short half-life: a structured feeling of lost polities and their dull affective orientations. Nonetheless, throughout this field of decaying propositions, there's something that sticks: a long-standing association between the erosion of truth and coercion.

"Coercion eclipses factuality" is hardly a novel proposition. At this point, its status seems closer to cliché. Whether or not "history is written by the victors" is the stuff of vigorous online debates.[13] Historical imaginaries spill over with revered figures (such as Socrates or Copernicus) who personify parrhesia,[14] countering the violent suppression of truth's pursuit as both a refrain throughout history, and a modality of history-making itself. Many thinkers have either cultivated, or critically questioned, a range of techniques — from seemingly benign, subtle manipulation to flat-out suppression — through which facts might be reshaped, eroded, distorted, or disappeared.[15] Orwell, of course, provided a clear diagnosis in his novel *Nineteen Eighty-Four* (1949).[16] Repeat after me: 2 + 2 = 5. Winston Smith, erstwhile employee at the Ministry of Truth, realigns his rationality according to the Ministry of Love's torturous new tune. Orwell voiced the threat of violence that lurks behind the knife's edge of state-sanctioned falsehoods, reordering even the most axiomatic and indisputable of mathematical truths.

Earlier, Edward Bernays had instrumentalised the suppler edges of rationality, reimagining public discourse according to desire's chaotic coursing, rather than rational, civic debate. Sigmund Freud's infamous nephew, who brought psychoanalysis to America and pioneered in public relations and propaganda, taught the twentieth century that consumer-citizens were subject to herd instinct and driven by passions more than logic. In an iconic early PR stunt, Bernays (commissioned by the American Tobacco Company) convinced more women to smoke by conflating cigarettes and women's liberation. He sent a float full of smoking suffragettes down Fifth Avenue in New York City's 1929 Easter Parade — a reordering of cigarettes's semantics that branded them as "torches of freedom" (psychoanalyst A. A. Brill's idea), and supposedly drove up sales across the country within weeks. Ironically, Bernays grossly and self-servingly exaggerated the extent of his own success with this campaign throughout subsequent decades of public lectures and unevidenced autobiographical writings — adding PR spin to PR tactics.[17] Also broadly and notably absent from Bernays's accounts of his success (especially so given the quasi-feminist trappings of his famous PR stunt) was the key role played by his wife, Doris E. Fleischman Bernays — his equal partner in the firm Edward L. Bernays, Counsel on Public Relations.[18] His PR spin on PR history yet again lends credence to his belief that public relations was not so much about promoting pre-constituted facts as it was "about fashioning and projecting credible renditions of reality itself".[19] His performance of said belief (in overstating his own success) demonstrates how PR carries the seeds of its own undoing — consolidating and undermining its claims to efficacy in a single gesture, through a series of ambivalently self-referential, performative speech acts and events.

In any case, Bernays cemented his reputation as a founding father of PR. By 1954, he had moved from advertising to politics, helping the CIA topple the democratically elected Guatemalan government.

Orwell paints a picture of flat-out violence corroding axiomatic truths. Bernays pioneers/*says* he pioneers the subtle arts of semantic realignment in the public sphere. Jean-François Lyotard and David Graeber, meanwhile, rethink the contact zones between contradictory social truths — and how those of one group might suppress, delegitimise or drown out those of another. Lyotard's concept of the "differend" encapsulates the lack of a universal judgment principle between two opposed but equally valid worldviews, in which case arriving at a sole judgment in a conflict situation would wrong at least one and possibly both parties.[20] Graeber (drawing from bell hooks and others)[21] thinks through colonial slavery and "interpretive labour" across racial, gender, and power divides. The masters, he notes, did not have to do much interpretive labour to understand their slaves' culture, thinking, or worldview. They had violence on their side. For slaves, on the other hand, interpretive labour was a highly foregrounded fact of life. Correctly interpreting a master's likes and dislikes, preferences and tastes could be a matter of life and death. Accompanied with the threat of violence, the master's minds and worldviews became objects of rich and nuanced interpretation, whereas the masters could completely overlook their slaves' worldviews — eroding their very claim to facticity.[22]

In the so-called "post-truth"[23] or "post-fact"[24] era, alliances between facticity and coercion have arguably changed shape at an accelerated pace. How so? This is a moment characterised by the circulation of hashtags, memes, and "fake news" — and of "fake news" thrown around as performative insult, by both vigilant publishing standards professionals, and sulking, power-hungry, would-be dictators. This is a moment characterised by coercive tactics woven deeply into myriad institutional and life practices, in an age of acute informatic and financial complexity (from Cambridge Analytica psyops influencing elections, to corporations' sneaky accounting procedures, designed to cheat workers out of pensions).[25] On the one hand, we could say that the coercion-factuality threshold has become more *personalised:* as covert data analysis operations gather pace, refining the idea of a *target* for political advertising, there is also a foregrounded emphasis on the figure of the gaslighting mastermind, pulling the wool over everyone's eyes. ("Donald Trump is Gaslighting America", reads one 2016 op-ed, which perfectly encapsulates this emphasis).[26] On the other hand, we might say that bullying has been *infrastructuralised,* seeping indistinguishably into ever-multiplying tactical fields. We might detect a hint of this sense in Nitzan and Bichler's 2009 account of capital as power (although they don't use the term bullying). The basis of capital, in their reading, is neither abstract labour (as in Marx), nor the util of neoclassical economics: it is power. Power, in turn, they define as *"confidence in obedience"* [...] "the certainty of the rulers in the submissiveness of the ruled".[27] More recently, Keller Easterling, the brilliant analyst of infrastructural dispositions, has addressed bullying in her account of "medium design" and the uselessness of being right in the current political landscape. She writes:

> *Oscillating between loops and binaries, an unnecessarily violent culture, having eliminated the very information it needs, is often banging away with the same blunt tools that are completely inadequate to address perennial problems and contemporary chemistries of power. [...] Since the world's big bullies and bulletproof forms of power thrive on this oscillation between loop and binary, it is as if there is nothing to counter them — only more ways of fighting and being right and providing the rancour that nourishes their violence.*[28]

"Common bullies and stubborn cross-purposes", for Easterling, "do not respond to reasonable solutions. They are even strange precipitates — or escapees — of those very attempts to tame the world with airtight logics".[29] Easterling's account of the bully as "strange precipitate" points to the possibility of developing an infrastructural reading of bullying. Such a reading could guide interventions for institutions that are both coercive and coerced, and within which bullying seeps beyond the figure of the bully, becoming a generalised disposition.

The figure of the bully

What is accomplished, discursively, by foregrounding the figure of the bully as exemplary of contemporary institutionality? What does the bully do — and what can it get away with? Space does not allow for a fully elaborated analysis of how bullying compares with a range of related phenomena such as coercion, abuse, harassment, manipulation, "strategic inefficiency",[30] epistemic injustice/epistemic violence,[31] cyberbullying,[32] trolling, workplace toxicity,[33] power-tripping, and gaslighting.[34] However, the account below will imply that bullying scenarios *can* include many of the above behaviours and phenomena, although said phenomena — by definition — may not necessarily constitute bullying. Though there is no consensus position (and, indeed, the term has undergone some surprising semantic shifts), for the purposes of my argument a provisional definition of institutional bullying might be this: *the use of coercive practices to reshape an institution (for example, to bypass dissenting views when introducing, evaluating, and deciding on policy shifts), often carried out by exerting pressure on colleagues' sense of being (via personal attacks, or reinforcing a sense of structural powerlessness), or their sense of being reliably oriented toward the institution's infrastructures.*

Already, this provisional definition (which differs from more standard definitions of bullying in its emphasis on how acts of bullying are directly imbricated in reshaping institutional policy) speaks to a certain closeness or proximity that typifies the relationships between bullies and institutions. The bully appears at the *zone of indistinguishability* between the shape of institutional policies and practices on the one hand, and workers' personal lives, affective lives, and senses of self on the other. This sense of closeness between the bully, the bullied, and the warp and weft of institutional decisions is interestingly illuminated by the etymological histories of bullying. Although today the connotations of bullying are clearly negative, "bully" initially appears to be derived from the Dutch *boel,* meaning "lover" or "brother"; in the sixteenth century, it meant "sweetheart". Throughout the seventeenth century, its meaning deteriorated: from "fine fellow" through to "harasser of the weak" by the 1680s, via the term *bully-ruffian.* An adjectival form, meaning "worthy, jolly, admirable" emerged in the 1680s and remained popular until the nineteenth century, preserving the earlier, laudatory sense of the word. The verb meaning "overbear with bluster or menaces" emerged in 1710.[35] Over time, the word shifts its senses of closeness, from endearing to menacing forms.

Recent writings on bullying largely focus on addressing and preventing bullying in workplaces and schools. For example, in the UK context, the non-departmental public body advising on employment relations, ACAS (Advisory, Conciliation and Arbitration Service), defines bullying as "offensive, intimidating, malicious or insulting behaviour, an abuse or misuse of power through means that undermine, denigrate or injure the recipient;" it lists examples such as "spreading malicious rumours, [...] exclusion or victimization, unfair treatment, overbearing supervision or other misuse of power or position, unwelcome sexual advances" and "deliberately undermining a competent worker by overloading and constant criticism".[36] According to UK employment law, bullying is not necessarily illegal, although harassment is; the latter can include bullying related to a protected characteristic as defined by the 2010 Equality Act (such as race, sex, age, disability, and pregnancy/maternity).[37] These senses of the term are certainly important, although they do little to interrogate the relationship between isolated *acts* of bullying and the very *shapes* of institutional policies and practices. Developing a picture of these complexities requires a rather less pragmatic approach to the problem of bullying.

While policy documents, counselling and self-help books on bullying abound, theoretical and philosophical approaches to bullying are harder to come by. One notable exception (alongside Easterling's texts above) is David Graeber's essay "The Bully's Pulpit'" (a clever twist on Theodore Roosevelt's 1904 phrase "the bully pulpit" to refer to the White House as a pleasing platform).[38] Graeber writes of schoolyard bullying as an "elementary structure" of domination — a situation that conditions both a widespread distaste for "sissies" of any kind, and the widespread conflation of bullies and "cowards" — such that the bullied seem just as reprehensible to people as do bullies.[39] In Graeber's reading, the schoolyard bully's authority is not at odds with the school's institutional authority; instead, "Bullying is more like a refraction of this authority", since, by mandating that pupils can't leave, institutions effectively hold victims in place for bullies.[40] Thus, Graeber counters the tendency for anti-bullying literature to either overlook the role of institutional authority in bullying scenarios, or assume that institutions play a benign role. The murky dynamics between bullies, victims, and witnesses create a scenario that Graeber terms the "'you two cut it out' fallacy", whereby *"Bullying creates a moral drama in which the manner of the victim's reaction to an act of aggression can be used as retrospective justification for the original act of aggression itself".*[41] The canny bully understands that, if his aggressions are pitched just right, the victim's response can be construed as the problem. Thus, for Graeber, the fundamental problem to which bullying points is not some mythologised "primordial aggressiveness" of the human species; rather, it is an inability to *respond* effectively to aggression: "Our first instinct when we observe unprovoked aggression is either to pretend it isn't happening or, if that becomes impossible, to equate attacker and victim, placing both under a kind of contagion, which, it is hoped, can be prevented from spreading to everybody else".[42]

Bullying takes root within institutions, we might say, by the very same process that makes the *figure* of the bully difficult to distinguish from the *ground* of "normal" institutional practices. The tendency for both the bully *and* the bullied to be seen as the problem leads to an ever-greater invisibility of bullying within the institution. A common response to workplace bullying is the decision not to report it, since it is often widely understood that HR departments' means of responding to complaints might be woefully under-nuanced. Such a response might even be (to paraphrase Sara Ahmed) "strategically inefficient" — so weak, delayed, or prolonged that it is at least as punitive for the complainant as for the accused.[43] Indeed, the most efficient response to workplace bullying might simply be to look away and shift one's career path (if possible) to dissociate oneself from the problem personality (or personality cluster). "Softer" institutional discourses such as gossip,[44] might pick up the windfall, fielding warnings about well-known bullies. Thus emerges the performative contradiction in the relationship between an institution and its bullies: because of the proclivity for institutions to *produce* such looking-away responses to bullying (based on a feeling that the institution *would* respond inadequately to a complaint), the very assumption that bullying acts according to a contagion-logic comes to be reinforced — such that, so to speak, the entire institution is *infected* by bullying — and it is not possible to separate the bullying "virus" from the institutional "host".

Graeber's account is brilliant, but my own account slightly reinterprets and refocuses the bully's relationship to institutional authority, shifting the emphasis away from schoolyard bullying and toward the adult world of the workplace — a context in which acts of bullying and institutional authority can be more directly imbricated. As a rule, pupils are not expected to make major contributions to schools' teaching and administration policies and practices. Some (though not all) colleagues, on the other hand, are expected to do so, to a greater or lesser extent, depending on the degree of leadership required by their roles. Thus, the figure of the workplace bully is one that emerges at the indecipherable edges of the institution as a *sedimentation of decisions* (the historically layered range of policies and practices that comprise it), and the institution as a *spectrum of personalities* — the figures who are (and/or who are seen as) the charismatic agents of particular decisions and policies.[45]

In the workplace, decisioning and bullying can be closely aligned. Insofar as an institutional decision is made by undermining staff personally until they drop a dissenting point of view and acquiesce to another staff member's decision, the shape of decision-making in the institution *is* the shape of bullying in the institution (To give one example: let's say a senior male staff member tells a junior female staff member that she is "taking this issue very personally" as an excuse to quickly override her objection to a particular policy decision. Formally, they are meant to find agreement across all parties in this situation; however, due to his seniority and better bargaining position with the senior management, he feels he has the upper hand in the negotiation and acts accordingly, feeling no particular need to entertain the logic of the dissenting view. Instead, his dismissal of the other staff member's "over-investment" acts as a shorthand to signal to everyone else in the room that the opposing idea is simply not going to happen. He's been acting like this for years, as is widely understood across the organisation). And yet, this shape of decision-making can never be straightforwardly interpreted as such, given that the range of "bullying" decisions (actioned with the aid of personal attacks, aimed at suppressing or pre-empting debate) may not be readily distinguishable from the non-bullying ones — except, perhaps, by a faint sense that a particular decision doesn't quite make sense. While the figure of the institution's bully barely surfaces (except, perhaps, at the edges of institutional discourse in gossip), the vague shape of its decisions can be taken as a forensic record, of sorts, to the bullying tides concocted, contained, and facilitated therein

— even if it cannot be "reverse-engineered" to reconstruct the power dynamics around the table.

Of course, some forms of bullying within institutions have nothing to do with setting out policy — or, for that matter, codes of professional practice. Institutions suppress some decisions and action others all the time, as a matter of course; this is entirely necessary for the institution to have anything close to a coherent set of practices. Many — perhaps even most — unactioned decisions might have been entirely unworkable in the first place. Further, institutions must operate within whatever unfavourable economic and policy contexts they might find themselves (as, for instance, when austerity measures "trickle down" to art institutions, making them more fiscally conservative). Even so, there is something very particular about collateral damage within the institutional decision-making scenario, justified or necessitated (so it might be argued by its perpetrators) by the need for speedy decision-making, and carried out via personal attack. The person whose objection — and therefore person — is construed as misguided, unjustified, or irrelevant, in becoming side-lined in the decision-making process, exemplifies an erosion of the distinctions between "personal", "affective", and "institutional" life that become active insofar as they enunciate a "weak point" in institutional procedure, where increased wilfulness (for better or worse) can easily reshape the institution. Bullying (whether tolerated *within* the institution or operating *as* the institution) cannot be easily identified through its forensic records as institutional decisions; but, perhaps, it can be *felt* that certain decisions take the shape of will-in-another-direction quashed — a *style of decisioning* that thrives on eroding the distinction between "private" and "institutional" life, and selects an appropriate aperture of witnessing to quickly propel the institution in the desired direction.[46]

The bully as anti-charismatic authority

Bullies craft witnessing situations within institutions to expand their wilfulness within them. More broadly, the figure of the bully has become foregrounded in its own right within recent political storytelling (carried out through news, blogs, and other online commentary), as a means to *stage* the dismantling of the institution for a wider audience. Bullies are imagined as slightly out-of-the-spotlight, but nonetheless powerful "back-of-house" decision makers providing the "quilting point", so to speak, that lightly tacks a prominent, public-facing, anti-institutional authoritarian to the institution and its own forms of authority. In this context, one can think, for instance, of how Leave-EU-campaign-manager-cum-Westminster-Chief-Advisor Dominic Cummings acted as a "behind the scenes" foil to UK Prime Minister Boris Johnson; or how campaign-strategist-cum-White-House-Chief-Strategist Steve Bannon has been construed as a puppeteer, of sorts, to US President Donald Trump.

Coursing prominently through news cycles, these bullying figures enact what I will call an *anti-charismatic* authority, which weds anti-institutional charismatic authority to institutional power. "Charismatic authority" is Max Weber's term for a type of authority wielded by compelling individuals, imbued with magnetism by passionate followers. Weber distinguishes charismatic authority from rational and traditional authority, and insists that the former is the very opposite of bureaucracy. Charisma stands fleetingly in relation to a proof of strength in *life,* rather than in established, abstract procedure;[47] fomented in the fervour of followers' devotion, and thus fleeting, unstable, and fundamentally opposed to the proceduralisation of power.[48] Thus emerges an elaborate set of problems as to how to make charismatic authority "stick" to a particular office or institution, beyond the gravitas of any one person who might have held that office. Weber recounts a range of succession rituals, which reckon with the problem of wedding charisma a bit more permanently to an office, transferring it from one, revered leader to (if all goes well) another.[49] Strategist-bullies like Cummings and Bannon, who back charismatic authoritarians like Johnson and Trump may well, indeed, have tried their own hands at gaining a following. Nonetheless, they really represent not charismatic authority as such, but *anti-charismatic authority:* rather than wedding charisma to an office through succession (as Weber describes), these figures provisionally tack volatile, anti-institutional, public-facing charismatic leaders to their offices, *translating leaders' professed anti-institutional attitudes into anti-institutional practices,* in an effort to maximise the institutional damage that charisma can inflict when repurposed as part of an institutional attack.

Take, for instance, Boris Johnson's former Chief Advisor, Dominic Cummings — an archetypal and much-remarked-on bully figure for the "post-truth" moment. Cummings has been widely denigrated as a bully in the press. (To cite one of the most theatrical, and indeed "witness-desiring" examples: in August 2019, Cummings sacked Tory chancellor Sajid Javid's media advisor, Sonia Khan, on suspicion of conspiring with anti-no-deal-Brexit Tories, without either proving the charge or consulting Javid about the dismissal, and had her marched out of No 10 by a police escort).[50] Cummings exemplifies bullying behaviour, professing zero tolerance for any range of opinion among Conservative ministers and parliamentarians that might compromise a hard-line, no-deal Brexit "negotiating position", and being seen as synonymous with the rise a "culture of fear"[51] in Westminster. He also exemplifies an intense hatred of bureaucracy in line with what Graeber has identified as a right-wing critique of the latter (namely: to understand the scourge of inefficient bureaucracy as a fundamental flaw of democracies, very much in contrast to the fabled efficiency of markets).[52] Cummings has expressed the desire to end the scourge of inefficient bureaucratic processes within government, drastically cutting both staff and "red tape".[53] His famously ruthless character has been used as a figurative shorthand for the anti-charismatic authority of institutional dismantling — called into question in a range of articles, talk shows, social media posts, television segments, and even a Channel 4 TV film called *Brexit: The Uncivil War* (2019). This latter — a prominent staging of the ruthless, right-wing campaign strategist that reckons with the lingering national trauma of the UK's 2016 EU referendum — featured Benedict Cumberbatch as a ruthless-yet-visionary Cummings, concocting a viable path for the Leave EU campaign's unlikely win. It features Cummings misdirecting left-behind voters' justified anger, and employing unprecedented micro-targeted, psychological voter manipulation via pioneering partnerships with shady data analytics firms. The figure of the bully moves fluidly "behind the scenes", from campaigning to government and back again: calling the shots; attacking psychological profiles and institutions at their weak points; shedding codes of conduct like so much collateral damage.

Is it any wonder that a figure like Cummings so neatly "figurates" both the anti-charismatic bully lodged in — attacking, infecting — the institution, *and,* indeed, the "post-truth" moment itself? The rhetorical task that figures like Cummings seem to accomplish is to package the thought that *post-truth is bullying:* a hatchet-man, lodged within the institution, attacking any soft, vulnerable, procedural edges that expose themselves to its spheres of contagion. The bully encapsulates exactly the hinge that links us, in our places of work and in our means to claim that we have been democratically *represented,* to the wider shift towards the anti-institutional, dismantling, obfuscating, plutocratic, divide-and-conquer procedures of the "post-truth" moment. This is a moment in which Brexit itself — as post-truth, micro-targeted, coercive, anger-misdirecting, disaster capitalist, racist-capitalist,[54] anti-bureaucratic, yet thoroughly bureaucratised fuck-up writ large — exemplifies the very inability of institutions such as parliaments to inoculate themselves against their bullies' attacks on institutional power.

Conclusion: Vice epistemologies

While, indeed, we may be witnessing a far-right desire to sabotage and dismantle institutions (perhaps, in the long run, only to replace these with other, as-yet nascent forms of authoritarian institutionality), the last thing I want to suggest is that this necessitates some wholesale turn away from institutional critique and its impulses — perhaps, along the lines of a nostalgic defence of institutions. Much to the contrary: perhaps nothing is more urgent than to rethink institutional practices. One way to do so would be to refocus institutional critique on the figure of the bully: a figure that seems to best typify the blurred lines between charisma and bureaucracy, racist and misogynist micro-aggressions and "business as usual", "life itself" and abstract proceduralism. From misogynistic dismissals of evidenced sexual harassment claims within offices, to ruthlessly efficient CEOs routinely under-staffing care facilities, and marching much loved line managers who fail to achieve criminally negligent budget-cut targets out of the building with security escorts, bullying abounds in institutions. Some such practices seem aimed at maintaining business as usual — preserving and fortifying fiefdoms within more-or-less established hierarchies. Others seem specifically (if not always directly) tied to budgetary discipline, and the demand to dismantle the institution's "inefficiencies". The face of these coercive practices — the bully — is partly "repurposed" as an austerity figure, restructuring the institution. Yet still, it remains ambiguous. Is the bully simply "tolerated" by the institution — or is bullying *the* institution? How does bullying align itself with other apparatuses of procedural change — or, conversely, oppressive stagnation — beyond the level of institutional governance? In a moment of endless puppeteering and pulling-the-strings, this ambiguity is arguably the bully's strength as a focus of analysis. Interpreted infrastructurally, the bully fruitfully exceeds the conceptual frame of the power-hungry "problem character". Instead, it speaks to the profoundly coercive nature of the so-called post-truth moment. Perhaps a focus on bullying might help institutional critique account for what, in business ethics, has recently been termed "vice epistemology":[55] the study of how epistemic vices (delusions, injustices, and other truth-eroding attitudes, characteristics, and dispositions) take hold within institutions, with an aim to remain "attentive to the context and conduct of individuals and groups operating in suboptimal epistemic conditions".[56] Starting with a clear-sighted appraisal of these suboptimal epistemic conditions — and the figures and forces that maintain them — might enable a response to institutional bullying that resists the urge to be "right", as Easterling would say,[57] and instead pays close attention to how bullying activates, or erodes the warp and weft of institutional procedures. This might enable new ways of thinking about bullying as a *tidal* force (so to speak) within institutions: never perfectly tied to particular figures or practices, but instead subject to rhythms of change as successive waves of management out-oppress, or better one another. Equally, thinking along these lines might energise discussions about what forms of collectivised decision-making can effectively inoculate institutions from bullying, and promote healthier epistemic environments in the process.

Notes

1. The term *hostile takeover* refers to one company taking over another against the will of the latter's management, either by proxy fights (coordinated shareholder attacks), or by tender offers (incitement of current shareholders to sell their shares), through which attempts to replace the management of the company to be acquired might be coordinated. In a moment in which certain key far-right political strategists, such as Steve Bannon, have come up through the Wall Street hostile takeover boom, I would like to suggest that the hostile takeover generalises as a mode of governance. For more on Bannon and the hostile takeover boom, see Joshua Green, *Devil's Bargain: Steve Bannon, Donald Trump, and the Storming of the Presidency* (London: Penguin, 2017).

2. SNAFU is military slang for System Normal: All Fucked Up.

3. To name just one example: UK prime minister Boris Johnson and his chief advisor Dominic Cummings's 2019 prorogation of parliament in an attempt to shove through a no-deal Brexit with minimal parliamentary scrutiny.

4. For more on the Windrush scandal, and the racialised

dimensions of UK immigration more broadly, see, for instance, Yasmin Gunaratnam, "On Researching Climates of Hostility and Weathering", in Kevin Smets *et. al*, eds, *The SAGE Handbook of Media and Migration* (London: SAGE, 2020), 129–141; Robbie Shilliam, *Race and the Undeserving Poor: From Abolition to Brexit* (Newcastle: Agenda Publishing, 2018); and Nisha Kapoor, *Deport, Deprive, Extradite: Twenty-first Century State Extremism* (London: Verso, 2018).

5. "Hostile environment" is Conservative, then Home Secretary Theresa May's 2012 term for a policy, through which it would be made increasingly difficult for those without leave to remain in the UK to stay in the country. See Lorenzo Pezzani, "Hostile Environments", *e-flux architecture*, 15 May 2020, accessed 2 August 2020, https://www.e-flux.com/architecture/at-the-border/325761/hostile-environments.

6. Naomi Klein, *The Shock Doctrine: The Rise of Disaster Capitalism* (London: Penguin, 2014).

7. Figures such as Crispin Odey perfectly typify the hedge fund manager speculating on Brexit. See, for instance, Jasper Jolly, "Hedge Funds make Big Bets against Post-Brexit UK Economy", *The Guardian*, 10 December 2018, accessed 2 August 2020, https://www.theguardian.com/business/2018/dec/10/hedge-funds-make-big-bets-against-post-brexit-uk-economy; and Lionel Laurent, "How the Hedge Funds are Really Playing Brexit", *Bloomberg Opinion*, 1 October 2019, accessed 2 August 2020, https://www.bloomberg.com/opinion/articles/2019-10-01/brexit-news-here-s-what-the-hedge-funds-are-really-up-to.

8. Intelligence and Security Committee of Parliament, *Russia*, 21 July 2020, accessed 2 August 2020, https://docs.google.com/a/independent.gov.uk/viewer?a=v&pid=sites&srcid=aW5kZXBlbmRlbnQuZ292LnVrfGlzY3xneDo1Y2RhMGEyN2Y3NjM0OWFl.

9. For an account of decisionism in the "post-truth" moment, see Luciana Parisi, "Reprogramming Decisionism", *e-flux* 85 (October 2017): https://www.e-flux.com/journal/85/155472/reprogramming-decisionism.

10. I am using the term "critique" here as a shorthand to point to the legacies of institutional critique; see, for instance, Biljana Ciric and Nikita Yingqian Cai, eds. *Active Withdrawals: Life and Death of Institutional Critique* (London: Black Dog Publishing, 2016); and Gerald Raunig and Gene Ray, eds., *Art and Contemporary Critical Practice: Reinventing Institutional Critique* (London: MayFly Books, 2009). However, it is important to note that the term critique itself has been questioned; for example, Irit Rogoff has described a shift in critical discourses from criticism to critique to criticality: "from finding fault, to examining the underlying assumptions that might allow something to appear as a convincing logic, to operating from an uncertain ground which, while building on critique, wants nevertheless to inhabit culture in a relation other than one of critical analysis". Irit Rogoff, "What Is A Theorist?" in *The State of Art Criticism*, ed. James Elkins and Michael Newman. (New York: Routledge, 2008), 99.

11. See Gilles Deleuze and Felix Guattari, *What is Philosophy?*, trans. Hugh Tomlinson and Graham Burchell (London: Verso, 1994), 61–83.

12. Alain Badiou, with Fabien Tarby, *Philosophy and the Event* (Hoboken: Wiley, 2013).

13. This aphorism is often attributed to Winston Churchill, although the attribution is questionable,

and versions of the phrase have appeared much earlier. Churchill also became one of the victors who wrote history; he wrote an account of World War Two based on his diaries, while others were forbidden to do so by the Official Secrets Act and won a Nobel Prize for Literature in 1953 for his trouble. Winston Churchill, *The Second World War* (London: Bloomsbury, 1959).

14. See Michel Foucault, *Fearless Speech,* ed. Joseph Pearson (Los Angeles: Semiotext(e), 2001).

15. I am especially thinking of techniques like Robert Thaler and Cass Sunstein's "nudge" theory within behavioural economics. Robert Thaler and Cass Sunstein, *Nudge: Improving Decisions about Health, Wealth, and Happiness* (New Haven: Yale University Press, 2008).

16. George Orwell, *Nineteen Eighty-Four: A Novel* (London: Arcturus Publishing, 2013).

17. Vanessa Murphree, "Edward Bernays's 1929 'Torches of Freedom' March: Myths and Historical Significance", *American Journalism* 32(3): 261.

18. See Susan Henry, "Anonymous in Her Own Name: Public Relations Pioneer Doris E. Fleischman", *Journalism History,* 23:2 (1997): 50–62, DOI: 10.1080/00947679.1997.12062467. I would like to thank Clea Bourne for drawing my attention to this point.

19. Stuart Ewen, *PR! A Social History of Spin* (New York: Basic Books, 1998), quoted in Murphree, "Edward Bernays's 1929 'Torches of Freedom' March", 260.

20. Jean-François Lyotard, *The Differend: Phrases in Dispute,* trans. Georges Van den Abbeele (Minneapolis: University of Minnesota Press, 1989), xi.

21. bell hooks, "Representations of Whiteness in the Black Imagination", in *Black Looks: Race and Representation* (New York: Routledge, 2015): 165–178.

22. David Graeber, "Dead Zones of the Imagination: On Violence, Bureaucracy, and Interpretive Labor", *HAU: Journal of Ethnographic Theory* 2, No. 2 (2012): 105–128.

23. See, for instance, Will Davies, "The Age of Post-Truth Politics", *New York Times,* 24 August 2016, accessed 2 November 2019, https://www.nytimes.com/2016/08/24/opinion/campaign-stops/the-age-of-post-truth-politics.html; Sheila Jasanoff and Hilton R. Simmet, "No Funeral Bells: Public Reason in a 'Post-Truth' Age", *Social Studies of Science* 47, No. 5 (2017): 751–770; Michael A. Peters, "Education in a Post-Truth World", *Educational Philosophy and Theory* 49, No. 6 (2017): 563–566; Sergio Sismondo, "Post-Truth?" *Social Studies of Science* 47, No. 1 (2017): 3–6; and Silvo Waisbord, "Truth is What Happens to News", *Journalism Studies* 19, No. 13 (2018): 1866–1878.

24. See, for instance, Peter Pomerantsev, *This is Not Propaganda: Adventures in the War Against Reality* (London: Faber & Faber, 2019), 163–167.

25. See, for instance, Ellen E. Schultz, *Retirement Heist: How Companies Plunder and Profit from the Nest Eggs of American Workers* (New York: Penguin, 2011).

26. Lauren Duca, "Donald Trump is Gaslighting America", *Teen Vogue,* 10 December 2016, accessed 28 July 2020, https://www.teenvogue.com/story/donald-trump-is-gaslighting-america.

27. Jonathan Nitzan and Shimshon Bichler, *Capital as Power: A Study of Order and Creorder* (London: Routledge, 2009), 17 [emphasis in original].

28. Keller Easterling, "Excerpt: Medium Design", *Strelka Mag,* 17 January 2018, accessed 28 July 2020, https://strelkamag.com/en/article/

keller-easterling-medium-design.

29. Keller Easterling, "Going Wrong", *e-flux architecture,* 10 December 2019, accessed 28 July 2020, https://www.e-flux.com/architecture/collectivity/304220/going-wrong.

30. Sara Ahmed, "Strategic Inefficiency", *Feminist Killjoys,* 20 December 2018, accessed 1 November 2019, https://feministkilljoys.com/2018/12/20/strategic-inefficiency.

31. See, for instance, Rachel McKinnon, "Gaslighting as Epistemic Violence: 'Allies,' Mobbing, and Complex Posttraumatic Stress Disorder, including a Case Study of Harassment of Transgender Women in Sport", In Benjamin R. Sherman and Stacey Goguen, eds., *Overcoming Epistemic Injustice: Social and Psychological Perspectives,* 285–302 (New York: Rowman & Littlefield, 2019).

32. See, for instance, Wendy Hui Kyong Chun, *Updating to Remain the Same: Habitual New Media* (Cambridge, MA: MIT Press, 2016), 135–165.

33. See, for instance, Joanna Wilde, *The Social Psychology of Organizations: Diagnosing Toxicity and Intervening in the Workplace* (London: Routledge, 2016).

34. See, for instance, Kate Abramson, "Turning up the Lights on Gaslighting", *Philosophical Perspectives* 28 (2014): 1–30; and Paige L. Sweet, "The Sociology of Gaslighting", *American Sociological Review* 84, No. 5 (2019): 851–875.

35. T.F. Hoad, ed., *The Concise Oxford Dictionary of English Etymology* (Oxford: Oxford University Press, 2003).

36. Advisory, Conciliation and Arbitration Service (ACAS), "Bullying and Harassment at Work: A Guide for Employees", accessed 4 November 2019, https://www.acas.org.uk/media/306/Advice-leaflet---Bullying-and-harassment-at-work-a-guide-for-employees/pdf/Bullying-and-harassment-at-work-a-guide-for-employees.pdf, 3–4.

37. GOV.UK, "Workplace Bullying and Harassment", accessed 4 November 2019, https://www.gov.uk/workplace-bullying-and-harassment.

38. David Graeber, "The Bully's Pulpit: On the Elementary Structure of Domination", *The Baffler* 28 (2015): 30–38, https://thebaffler.com/salvos/bullys-pulpit.

39. Graeber, "The Bully's Pulpit".

40. Graeber, "The Bully's Pulpit".

41. Graeber, "The Bully's Pulpit", [emphasis in original].

42. Graeber, "The Bully's Pulpit".

43. Ahmed, "Strategic Inefficiency".

44. For one detailed account of the role of gossip and rumour, see Gavin Butt, *Between You and Me: Queer Disclosures in the New York Art World, 1948–1963* (Durham: Duke University Press, 2005).

45. See Max Weber, *On Charisma and Institution Building,* ed. S. N. Eisenstadt (Chicago: University of Chicago Press, 1968).

46. Although it is beyond the scope of this essay to develop this point further, I intend this point as an extension of previous debates on emotional labour in workplaces within the austerity-era institution. For a ground-breaking early account of emotional labour in the workplace, see Arlie Russel Hochschild, *The Managed Heart: Commercialization of Human Feeling* (Berkeley: University of California Press, 1983).

47. Weber writes, "The charismatic leader gains and maintains authority solely by proving his strength in life", and speaks of charisma as having a "way of life flowing from its meaning". Weber, *On Charisma and Institution Building,* 22.

48. Weber writes, "charismatic domination is the very opposite of

bureaucratic domination". He insists that charisma "does not embrace permanent institutions", or, indeed, ordered economy: "it is the very force that disregards economy". Weber, *On Charisma and Institution Building,* 20–21.

49. Weber, *On Charisma and Institution Building* 54–61.

50. Kate Proctor, "Sajid Javid was Not Told in Advance of Adviser's Sacking by Cummings", *The Guardian,* 30 August 2019, accessed 30 November 2019, https://www.theguardian.com/politics/2019/aug/30/sajid-javid-was-not-told-in-advance-of-advisers-sacking-by-cummings.

51. Kate Proctor, "'Culture of Fear' Claims as Javid Confronts PM Over Adviser's Sacking", *The Guardian,* 30 August 2019, accessed 1 December 2019, https://www.theguardian.com/politics/2019/aug/30/sajid-javid-confronts-boris-johnson-over-advisers-sacking.

52. David Graeber, *The Utopia of Rules: On Technology, Stupidity, and the Secret Joys of Bureaucracy* (Brooklyn: Melville House, 2015), 7.

53. This desire to rid governments of checks, balances, and red tape is reminiscent of some signs of ruthless efficiency deployed during the early days of the US Trump presidency (behind the veil of crazy tweets), such as the chilling one-sentence bill H.R. 861, calling for the termination of the Environmental Protection Agency as of 31 December 2018. In spite of its evincing the ruthless trashing of checks and balances, this bill was widely viewed as merely a "messaging bill" – designed to attract attention, even though it clearly had no chance of getting through Congress. Justin Worland, "No, President Trump Isn't Going to Eliminate the EPA. But He Might Do This", *Time,* 16 February 2017, accessed 1 December 2019, https://time.com/4673233/epa-elimination-donald-trump-scott-pruitt.

54. Although it is beyond the scope of this essay, much more could be said about the immense significance of analyses of racial capitalism in this political moment. See Cedric J. Robinson, *Black Marxism: The Making of the Black Radical Tradition* (Chapel Hill: University of North Carolina Press, 2000); and for a more recent account, Gargi Bhattacharyya, *Rethinking Racial Capitalism: Questions of Reproduction and Survival* (Lanham: Rowman & Littlefield International, 2017).

55. Christopher Baird and Thomas S. Calvard, "Epistemic Vices in Organizations: Knowledge, Truth, and Unethical Conduct", *Journal of Business Ethics* 160 (2019): 263–276.

56. Baird and Calvard, "Epistemic Vices in Organizations", 263–276.

57. Keller Easterling, "Excerpt: Medium Design".

References

Abramson, Kate. "Turning up the Lights on Gaslighting". *Philosophical Perspectives* 28 (2014): 1–30.

Advisory, Conciliation and Arbitration Service (ACAS). "Bullying and Harassment at Work: A Guide for Employees". Accessed 4 November 2019. https://www.acas.org.uk/media/306/Advice-leaflet---Bullying-and-harassment-at-work-a-guide-for-employees/pdf/Bullying-and-harassment-at-work-a-guide-for-employees.pdf.

Ahmed, Sara. "Strategic Inefficiency". *Feminist Killjoys,* 20 December 2018. Accessed 1 November 2019. https://feministkilljoys.com/2018/12/20/strategic-inefficiency.

Badiou, Alain, with Fabien Tarby. *Philosophy and the Event.* Hoboken: Wiley, 2013.

Baird, Christopher and Thomas S. Calvard. "Epistemic Vices in

Organizations: Knowledge, Truth, and Unethical Conduct". *Journal of Business Ethics* 160 (2019): 263–276. https://doi.org/10.1007/s10551-018-3897-z.

Bhattacharyya, Gargi. *Rethinking Racial Capitalism: Questions of Reproduction and Survival.* Lanham: Rowman & Littlefield International, 2017.

Brexit: The Uncivil War. Dir. Toby Haynes. Baffin Media, 2019.

Butt, Gavin. *Between You and Me: Queer Disclosures in the New York Art World, 1948–1963.* Durham: Duke University Press, 2005.

Chun, Wendy Hui Kyong. *Updating to Remain the Same: Habitual New Media.* Cambridge, MA: The MIT Press, 2016.

Churchill, Winston. *The Second World War.* London: Bloomsbury, 1959.

Ciric, Biljana and Nikita Yingqian Cai, eds. *Active Withdrawals: Life and Death of Institutional Critique.* London: Black Dog Publishing, 2016.

Davies, Will. "The Age of Post-Truth Politics". *New York Times,* 24 August 2016. Accessed 2 November 2019. https://www.nytimes.com/2016/08/24/opinion/campaign-stops/the-age-of-post-truth-politics.

Deleuze, Gilles and Felix Guattari. *What is Philosophy?* Translated by Hugh Tomlinson and Graham Burchell. London: Verso, 1994.

Duca, Lauren. "Donald Trump is Gaslighting America". *Teen Vogue,* 10 December 2016. Accessed 28 July 2020. https://www.teenvogue.com/story/donald-trump-is-gaslighting-america.

Easterling, Keller. "Excerpt: Medium Design". *Strelka Mag,* 17 January 2018. Accessed 28 July 2020. https://strelkamag.com/en/article/keller-easterling-medium-design.

Easterling, Keller. "Going Wrong". *e-flux architecture,* 10 December 2019. Accessed 28 July 2020. https://www.e-flux.com/architecture/collectivity/304220/going-wrong.

Elkins, James and Michael Newman, eds. *The State of Art Criticism.* New York: Routledge, 2008.

Ewen, Stuart. *PR! A Social History of Spin.* New York: Basic Books, 1998.

Foucault, Michel. *Fearless Speech.* Edited by Joseph Pearson. Los Angeles: Semiotext(e), 2001.

GOV.UK. "Workplace Bullying and Harassment". Accessed 4 November 2019. https://www.gov.uk/workplace-bullying-and-harassment.

Graeber, David. "Dead Zones of the Imagination: On Violence, Bureaucracy, and Interpretive Labor". *HAU: Journal of Ethnographic Theory* 2, No. 2 (2012): 105–128.

Graeber, David. "The Bully's Pulpit: On the Elementary Structure of Domination". *The Baffler* 28 (2015): 30–38. https://thebaffler.com/salvos/bullys-pulpit.

Graeber, David. *The Utopia of Rules: On Technology, Stupidity, and the Secret Joys of Bureaucracy.* Brooklyn: Melville House, 2015.

Green, Joshua. *Devil's Bargain: Steve Bannon, Donald Trump, and the Storming of the Presidency.* London: Penguin, 2017.

Henry, Susan. "Anonymous in Her Own Name: Public Relations Pioneer Doris E. Fleischman". *Journalism History* 23:2 (1997): 50–62. DOI: 10.1080/00947679.1997.12062467.

Hoad, T.F., ed. *The Concise Oxford Dictionary of English Etymology.* Oxford: Oxford University Press, 2003.

Hochschild, Arlie Russel. *The Managed Heart: Commercialization of Human Feeling.* Berkeley: University of California Press, 1983.

hooks, bell. "Representations of Whiteness in the Black Imagination". In *Black Looks: Race and Representation.* New York: Routledge, 2015: 165–178.

Intelligence and Security Committee of Parliament, Russia. 21 July 2020. Accessed 2 August 2020. https://docs.google.com/a/independent.gov.uk/viewer?a=v&pid=sites&srcid=aW5kZXBlbmRlbnQuZ292LnVrfGlzY3xneDo1Y2RhMGEyN2Y3NjM0OWFl.

Jasanoff, Sheila and Hilton R. Simmet. "No Funeral Bells: Public Reason in a 'Post-Truth' Age". *Social Studies of Science* 47, No. 5 (2017): 751–770.

Jolly, Jasper. "Hedge Funds make Big Bets against Post-Brexit UK Economy". *The Guardian,* 10 December 2018. Accessed 2 August 2020. https://www.theguardian.com/business/2018/dec/10/hedge-funds-make-big-bets-against-post-brexit-uk-economy.

Kapoor, Nisha. *Deport, Deprive, Extradite: Twenty-first Century State Extremism.* London: Verso, 2018.

Klein, Naomi. *The Shock Doctrine: The Rise of Disaster Capitalism.* London: Penguin, 2014.

Laurent, Lionel. "How the Hedge Funds are Really Playing Brexit". *Bloomberg Opinion,* 1 October 2019. Accessed 2 August 2020. https://www.bloomberg.com/opinion/articles/2019-10-01/brexit-news-here-s-what-the-hedge-funds-are-really-up-to.

Lyotard, Jean-François. *The Differend: Phrases in Dispute.* Translated by Georges Van den Abbeele. Minneapolis: University of Minnesota Press, 1989.

McKinnon, Rachel. "Gaslighting as Epistemic Violence: 'Allies,' Mobbing, and Complex Posttraumatic Stress Disorder, Including a Case Study of Harassment of Transgender Women in Sport". In Benjamin R. Sherman and Stacey Goguen, eds. *Overcoming Epistemic Injustice: Social and Psychological Perspectives,* 285–302. New York: Rowman & Littlefield, 2019.

Murphree, Vanessa. "Edward Bernays's 1929 'Torches of Freedom' March: Myths and Historical Significance". *American Journalism* 32, no. 3 (2015): 258–281.

Nitzan, Jonathan and Shimshon Bichler. *Capital as Power: A Study of Order and Creorder.* Abingdon: Routledge, 2009.

Orwell, George. *Nineteen Eighty-Four: A Novel.* London: Arcturus Publishing, 2013.

Parisi, Luciana. "Reprogramming Decisionism". *e-flux* 85 (October 2017). https://www.e-flux.com/journal/85/155472/reprogramming-decisionism.

Peters, Michael A. "Education in a Post-Truth World". *Educational Philosophy and Theory* 49, No. 6 (2017): 563–566.

Pezzani, Lorenzo. "Hostile Environments". *e-flux architecture,* 15 May 2020. Accessed 2 August 2020. https://www.e-flux.com/architecture/at-the-border/325761/hostile-environments.

Pomerantsev, Peter. *This is Not Propaganda: Adventures in the War Against Reality.* London: Faber & Faber, 2019.

Proctor, Kate. "'Culture of Fear' Claims as Javid Confronts PM Over Adviser's Sacking". *The Guardian,* 30 August 2019. Accessed 1 December 2019. https://www.theguardian.com/politics/2019/aug/30/sajid-javid-confronts-boris-johnson-over-advisers-sacking.

Proctor, Kate. "Sajid Javid was Not Told in Advance of Adviser's Sacking by Cummings". *The Guardian,* 30 August 2019, Accessed 30 November 2019. https://www.theguardian.com/

politics/2019/aug/30/sajid-javid-was-not-told-in-advance-of-advisers-sacking-by-cummings.

Raunig, Gerald and Gene Ray, eds. *Art and Contemporary Critical Practice: Reinventing Institutional Critique*. London: MayFly Books, 2009.

Robinson, Cedric J. *Black Marxism: The Making of the Black Radical Tradition.* Chapel Hill: University of North Carolina Press, 2000.

Rose, Jonathan. "Brexit, Trump, and Post-Truth Politics". *Public Integrity* 19, No. 6 (2017): 555–558.

Schultz, Ellen E. *Retirement Heist: How Companies Plunder and Profit from the Nest Eggs of American Workers.* New York: Penguin, 2011.

Shilliam, Robbie. *Race and the Undeserving Poor: From Abolition to Brexit.* Newcastle: Agenda Publishing, 2018.

Sismondo, Sergio. "Post-Truth?" *Social Studies of Science* 47, No. 1 (2017): 3–6.

Smets, Kevin Koen Leurs, Myria Georgiou, Saskia Witteborn and Radhika Gajjala, eds. *The SAGE Handbook of Media and Migration.* London: SAGE, 2020.

Sweet, Paige L. "The Sociology of Gaslighting" *American Sociological Review* 84, No. 5 (2019): 851–875.

Thaler, Robert and Cass Sunstein. *Nudge: Improving Decisions about Health, Wealth, and Happiness.* New Haven: Yale University Press, 2008.

Waisbord, Silvio. "Truth is What Happens to News". *Journalism Studies* 19, No. 13 (2018): 1866–1878.

Weber, Max. *On Charisma and Institution Building.* Edited by S. N. Eisenstadt. Chicago: University of Chicago Press, 1968.

Wilde, Joanna. *The Social Psy-chology of Organizations: Diagnosing Toxicity and Intervening in the Workplace.* London: Routledge, 2016.

Worland, Justin. "No, President Trump Isn't Going to Eliminate the EPA. But He Might Do This". *Time,* 16 February 2017. Accessed 1 December 2019. https://time.com/4673233/epa-elimination-donald-trump-scott-pruitt.

Trust without Evidence: Post-Truth, Culture Policies and Funding Dependency

Carolina Rito

I stumble over the expression "post-truth", struck by the promise it bears on a rather forthright past. The idea that we lived in a time when institutions and media outlets were held accountable for the veracity and honesty of their statements is something comforting only to imagine. The phrase post-truth arguably holds the promise of a monolithic reality, equally accessible to all, regardless of origin, nationality, gender, socio-economic background, and skin colour. We recognise that there has been a shift in patterns of news production and consumption since the 2008 financial crisis and the increased use of social media platforms. However, it is worth reflecting on the suitability of the term post-truth to convey that very same shift and the promises it holds. This text draws on the contemporary longing for truth, in order to explore how it manifests within the neoliberal paradigm. I will argue that this paradigm has driven a system based on the economy of evidence (to uphold the veracity of facts). I aim to explore how our trust in evidence was turned into a bureaucratic ruse — i.e. dependency on external funding — that governs the function of cultural practitioners and institutions, and holds to a prescribed future.

The NSA files and The Guardian incident

When reflecting on the contemporary paradigm of media outlets, an image that comes to mind is the staff at British newspaper *The Guardian,* in the basement of their London headquarters, destroying computers used to store top-secret documents leaked by the US National Security Agency whistle-blower Edward Snowden in 2014. This destruction was directly ordered by David Cameron, the then Conservative Prime Minister of the UK, and was witnessed and recorded by technical experts of the British intelligence and security organisation Government Communications Headquarters (GCHQ).[1] This case is complex — not to mention too hilarious to be true — from a technical and philosophical perspective. For instance, the hard work involved in tearing apart a computer and in drilling thick metal pieces; and the questions that emerge around the materiality of data and software as holders of evidence. I won't attempt a comprehensive analysis here, however appealing. Rather, there are a couple of things that are worth stressing: the disheartening, direct, and ruthless influence of democratic states in the (so-called) affairs of the free press, and — not as depressing but definitely more telling to my argument — the public trust in the evidence. Trust is not only a personal belief in something. It also implies, in legal terms, a contract of management and exploitation whereby something is put forward to the possession of someone else (the trustee) to be held or administered for the benefit of another.[2] In this contractual link between trustee and beneficiary, political theorist Angela Mitropoulos identifies some of the principles of the construction of "legitimated forms of subjectivity and relation that have accompanied the rise and expansion of capitalism across the world".[3]

Returning to the scene in the basement of *The Guardian,* I cannot help but linger on the fact that the two GCHQ officers that commanded the operation recorded the effort on their phones. The recording happened after they, and the institution they served, had been reassured by *The Guardian* editor Alan Rusbridger that the destruction of the leaked files would not stop the intelligence-related news, since there were more copies, at least in Brazil and the US.[4] If the footage was not meant to prove the evidence-destruction of the misdeeds of democratic countries and prevent the spread of such news, then, why bother to record the destruction, or even destroy it in the first place? The impetus to record is of crucial importance. It seems to represent a fundamental belief or trust in the document, held by these two officers, as the support that bears the evidence. Somehow, it is a respectful conviction to which, certainly, a lot of us relate. However, the bottom line is that the need to record reveals the capacity to hold two completely contradictory beliefs in one's mind at the same time, accepting them as equally valid; what George Orwell called "doublethink" in his influential book *Nineteen Eighty-Four* (1949).[5] This double-bind belief is reflected in the fact that the physical destruction of the computers' hard drives did not represent the obliteration of the evidence. The role of the document overlaps with the function of the contract, bounding both parties together, one subjugated to the other. Moreover, the impossible obliteration of the evidence co-habits with its opposite, that is, the officers' trust in the audio-visual documentation of the destruction of evidence. Or, as Jean Baudrillard argued in his series of articles published in the *Libération* during the Gulf War, the audio-visual representation — the recording — displaces the experience of the event to the seeming veracity of the document and its ambition to bear evidence.[6]

However, the NSA files incident at *The Guardian* office is not only paradigmatic evidence of our mistrust in the phrase "post-truth", but also of how the expression seems to hold a certain nostalgia for real information, implying that a desirable regime lies in the past. Trusting the evidence also suggests forgetting the lessons of the linguistic turn and post-structuralist thinking, whereby we learnt that documents (language-based or otherwise) are necessarily socio-cultural and subjective articulations. Visual culture theorist Ariella Aïsha Azoulay reminds us that what we sometimes take to be ideal systems, such as democracy, or, for the purpose of this text, the era of reliable information, are rather optimistic and naïve beliefs.[7] For Azoulay, the foundations of democracy are a stronghold of persistent violence and the subordination of others, with greater stress on its colonial iterations. In a similar way, in the Enlightenment, the idea of a universal truth was owned and deployed by a few, who sometimes via force, sometimes via soft power mechanisms (such as scientific knowledge) created the belief that there was such thing as a univocal worldview to legitimise the disposability of everything that fell outside the hegemonic framework. Following Foucault's analysis of the mechanisms of power and their relation to truth-making, this framework is a situated, time-specific regime that it is worth analysing and should not be taken for granted.[8]

Evidence in the aftermath of the Covid-19 pandemic

Evidence has gained yet another new value in contemporary times. I am writing two months into the announcement of the Covid-19 pandemic lockdown in the UK.[9] We are, arguably, more knowledgeable about virology, epidemiology, prognostication of cases and death toll, apocalyptic theories, and well-being tips for working from home. We follow the national news and check in with the Johns Hopkins University of Medicine's daily summary of new cases and fatalities, which collects data from across the world.[10] We read analytical readings of current data, and prognostic articles projecting what the new normal might look like. Simultaneously, our screens and everyday online conversations are inundated with the surge of fake news, do-it-yourself cures, conspiracy theories, and optimistic prognostics for a new vaccine. It is within this context that Billy Palmer, contributor to the Nuffield Trust and Senior Fellow in Health Policy in the UK, noted in an article three weeks after the lockdown announcement that:

> *Given the understandable clamour for informative data on the pressures the pandemic is placing on health care and the effectiveness of the response in different countries, there is a risk that if authorities fail to provide sufficient information this will create a vacuum to be filled by sensationalist, inaccurate or wholly fake news.*[11]

Palmer calls for more official data to avoid the surge in fake news. Although we hear Palmer's concerns, it is now evident that authorities' numbers have been the greater contributors to the inflation of this information vacuum. Palmer's claim for sufficient information does not take into account the variable that numbers mean very little if not accompanied by the formula and criteria that have generated them. For instance, the UK reporting of fatalities did not include deaths of residents in care homes until early May (two months after the first reported deaths in the country), or any victims who died outside of national hospitals. Additionally, post-mortem tests in non-diagnosed Covid-19 patients were not conducted. Although apparently simple to calculate, the numbers are treated differently, and, as Tiago Marques, Associate Professor in Biology at Lisbon University, reminds us, when reflecting on the Covid-19 numbers published by national authorities, these fail to represent the reality:

> *We cannot trust the numbers as if they were an absolute truth, because naturally these numbers correspond to observations, which can differ from reality. We would like to observe reality without error, however, unfortunately, reality is almost always inaccessible.*[12]

Trust in the evidence and the cultural sector

Despite the effort to achieve more effective evidence in the sciences and politics, journalism and history, it is with a splintered conception of reality and, in fact, contributing to its dissolution, that the arts have critically addressed the "neo-positivist conception of evidence".[13] Echoing Marques's view, the arts have long been developing a critical inquiry that not only observes the mechanisms of truth-making that govern us, but also sets up a line of inquiry about belief systems. I would like to reflect on the ways in which our trust in evidence and its legal apparatus has affected the cultural sector, and, moreover, how it has impacted the purpose of contemporary art practitioners and its institutions. For this, we need to go back to 2008.

The economic crisis of 2008 has affected the European welfare infrastructure in unprecedented ways since the Second World War, giving rise to the acceleration of neoliberal policies across all areas of society. The sense of urgency galvanised by the media made the austerity measures imposed by the financial sector look inevitable. These measures were implemented overnight, with the permission and complicity of the sector's political allies. The International Monetary Fund widely intervened in countries' internal governance, forcing the pervasive privatisation of national resources and protecting the banks — which was said to be crucial to keep the economy from collapsing. At the same time, the cultural sector was being "redesigned" to serve new agendas. Cultural practitioners found themselves with no other option but to work for the cultural industries, whereas public institutions saw their funding reduced dramatically. Expected to continue delivering the same cultural services, but now with less resources, institutions dismissed many of their staff, especially those who were already in precarious employment contracts. Simultaneously, new informal jobs were created under "zero-hours contracts" — proving that these posts were still vital to the functioning of many institutions.

While public funding was being squeezed and the hesitant infrastructures of the welfare state being repurposed for private profit, cultural institutions were asked to diversify their portfolio towards financial sustainability. Arguably, this would be the only way to prevent institutions from closing their doors to their audiences, since public money was not secured anymore. It meant that exhibition galleries, and multi-purpose spaces (otherwise available for programming, and to practitioners and local communities to meet and work) were now being offered for commercial hire. Cafés opened in public cultural institutions or got rebranded in a fancier and more expensive fashion. Museums expanded their shops, whereas those without one created them, usually occupying a prevalent space between the entrance and the reception — see, for instance, Nottingham Contemporary and the Institute of Contemporary Arts (ICA), London, in the UK.

Another important staple of the new financial profile of cultural organisations was to diversify their sources of funding, moving from a majority of public funding to a rapid increase of private income streams. To serve this purpose, the teams and the focus of their expertise changed, with a reduction in museum experts such as curators, and a directly inverse growth of development and marketing teams. In a recent interview, the Director of the ICA in London, Stefan Kalmár, analyses the transformation of the sector from a welfare state approach to a neoliberal paradigm:

> *[L]ike the NHS, we were founded in the Keynesian economic aftermath of the Second World War, recognising the need for essential services including healthcare, social welfare, housing, transport, education, and culture and media to be public services and a counterweight to those of capitalism. Subsequently, with the Milton Friedman school of thought, this went out of the window in the 1970s. The ICA today receives only 21% public funding — so, strictly speaking, the public is not even any longer a majority stakeholder — and let's not forget, we are talking here about an iconic British institution that invented the very idea of an institute of contemporary arts, renowned around the world as a think tank of tomorrow and the birthplace of pop art.*
>
> *So, what does this all mean? It means that over the past 40 years, cultural organisations have been forced to operate more and more commercially. We've been forced to "diversify our income streams" — a bit retail, a bit ticket sales, a bit individual giving, a bit sponsorship, food and beverage, editions and so on. It's inherently a precarious economy, actually often not that dissimilar to the precarious economic reality of many artists themselves (bar jobs, art sales, teaching, writing).*[14]

Kalmár leaves us with a clear picture of the contemporary paradigm of cultural institutions and how the logics of neoliberalism were implemented in the sector in a one-size-fits-all style. The impact of the neoliberalisation of the sector has been discussed elsewhere in greater detail, so I won't attempt another analysis here.[15] Nevertheless, I would like to look at this fund-raising activity as an extension and mutation of the "trust in the evidence" and how this neo-positivist approach, ultimately and consequently, represents an obstacle to the activity that these institutions were meant to serve: accommodating the ever-changing landscape of the arts and its modes of engagement. As I have argued above, trust is not only a personal belief. Rather, it creates a bond between the trusted and the trustees, which, in the case of fund-raising activities, is established between the applicant and the funding-body. The external/private funding-dependency relegates the sector's agenda and priorities to the hands of a few: those responsible for defining the financial strategy of the funding calls. Once more, as in the case of *The Guardian* headquarters footage, the evidence — or the funded project brief — becomes a legally binding agreement that determines the future and scope of action of the applicant.

In practical terms, funding application forms are populated with sections for aims, objectives, outputs, outcomes, beneficiaries, audiences, and impact. And eventually complemented with some other sections; for example, explain how your proposal responds to the call priorities, or the funding-body strategies, or the ethos of this year's award. Arguably, of the seven sections listed above, at least five are impossible to anticipate; moreover, even if one can roughly identify some of the aforementioned aspects, these are areas that should remain open in the context of cultural activities. Contrary to the neoliberal pursuit of anticipation and inscription of what cultural practices can enable, cultural activity should be driven by curiosity and an exploratory and experimental approach.

How does the ambition for objectivity in the funding dependency of the cultural sector speak to the faith in evidence demonstrated by the GCHQ officers in the NSA files leak described above? These two cases meet in the implementation of a temporal binding reified in the document — be it the footage, or the successful application brief. The recording of the destruction of hard drives with leaked NSA files, and funding application briefs are binding contracts in the present, between the parties involved, defining a future to come. Although the footage does not destroy the possibilities of publishing news on the scandal, it is clear that the target newspaper is signing an agreement of non-publication of any related material, arguably bound by the evidence of the event in the basement. The documentation, in its contract capacity, compromises the potential of a polyvocal world, and the accountability of the states for the crimes committed. In this way, what is captured is not only our present time — filling in an application form and finding ways to engage with its objective sections — but the nature of what can be programmed. Application forms operate as binding scripts for that future to come; an anachronic prediction.

Media theorist Wendy Hui Kyong Chun argues that these projections are not only predictions, but that they also have the power to determine and inform the future. Although Chun writes within the context of big data and machine-learning, an extrapolation to the realm of fundraising and cultural funding policies can be considered here, inasmuch as what is being observed is the paradigm of the apparatus of future prediction. Paradigms are not compartmentalised in disciplines; instead, they traverse them in a continuum. Thus, according to Chun:

> *A lot of the hype around big data and a lot of machine-learning programs stems from their alleged predictive power. Basically, they argue that "based on the past, we can predict the future". But not only do they predict the future, they often put the future in place. Their predictions are correct because they program the future. [...] Think of something like a risk-management system for credit. They'll do a risk assessment of your credit based on your education, social networks, etc., and then they'll give you credit or not — or give you credit at a certain interest rate. In effect, by denying you credit, they're affecting what your future will be. [...] These predictions are treated as truth and then acted upon.*[16]

As with risk-management systems for credit, funding call requirements in the cultural sector are based on the past, that is to say, on the tradition of aesthetics, on previous and tested formats, on existing audience segments, etc. Equally, the funding apparatus causes that known and tested framework to be rep-eated in the future, securing the governance of what can be programmed, curated, and, ultimately, experienced. To ensure that the cycle is brought home, the outputs and outcomes will have to be demonstrated in a final report to the competent bodies.

The predictive apparatus takes over the institution's planning, resources, and programming, whereas its predictive power subordinates the artistic and curatorial practices with incommensurable consequences. They reach beyond the mere logistics of finance. The proceedings affect the operation of critical practices by curtailing, in advance and based on the past, these practices' potential. As I have written elsewhere, contemporary art institutions have moved away from an exclusive collecting, conservation, and display remit to accommodate a more flexible approach, adequate to house the ever-occurring transformations happening in the field.[17] Often in Kunsthalle-like institutions, the remit is to respond to the formats and conceptual developments of the provocative and exploratory practices engaging with polyvocal modus operandi, worldviews, and wider societal issues.

Neoliberal cultural utilitarianism

In addition to concerns around the prescriptive power of funding dependency, it is important to introduce another element to the equation: the rebranding of the role of the arts in the context of neoliberal cultural policies, or what I call the ideology of cultural utilitarianism. By that I mean the ideology under which culture is seen as a service provider for concrete results in society at large; for example, tackle societal diversity and equality, elders' loneliness, and young people's school attainment. Privatisation and the neoliberalisation of cultural institutions — along with the funding apparatus — come hand-in-hand with funding cuts and the outsourcing of key welfare services. After Thatcher's infamous statement that "there is no such thing as society", David Cameron responded with the conservative rebranding slogan Big Society, which has mutated into an outsourcing of debt that is now bound to the capitalist recovery of the conditions of gratuitous labour. This new approach applies the political ideology of the free market to the functioning of society dynamics. According to Corbett and Walker:

> *The "Big Society" draws on a mix of conservative communitarianism and libertarian paternalism. Together, they constitute a long-term vision of integrating the free market with a theory of social solidarity based on hierarchy and voluntarism.*[18]

The Conservatives' Big Society comes to reinvent a new neoliberal conception of community; one that leaves society at its own mercy, intentionally mistaken for the idea of freedom. Appropriately, it supplements key social and cultural services, while encouraging socio-economic inequalities, hate crime, xenophobia, mental health issues, and the disempowerment of the most vulnerable. According to the neoliberal vision for art practices and cultural institutions, these ever-growing vulnerable infrastructures are also called on to repair the after-effects of the crisis. As Tazzioli and Lorenzini rightly point out, the "'trap of presentism' [...] is at the core of problem-solving analyses and imposes on us a specific and monolithic temporality — one that is often conceived in terms of 'crisis'".[9] A good example are the ubiquitous funding streams addressing the challenges of Covid-19 over the past two months (and that are expected to continue for at least another year), which leaves any other research topic/urgency underfunded, thus channelling resources — regardless of the field, discipline, or practice — to the new "crisis".

Underfunded, understaffed, and worried about survival, professionals in the cultural sector now also feel the urge to repurpose their skills to help "fix the mess". Don't get me wrong here. I am not against the complementarity of approaches to tackle societal issues, for which the arts can be of great inspiration and use. What I am trying to point out is the impoverishment of key welfare services, reliant on professional and highly skilled labour, which are then replaced with the (low-cost) non-specialised and "creative" response of cultural practitioners. Keeping cultural practitioners on their toes, inviting them to react to yet another brand-new crisis, while precarity mounts and labour stability is being pulled out from under their feet.[20]

Final notes

In the cultural sector, the decisions about what to address, programme, and, ultimately, where to invest (in terms of funding, but also in terms of time and resources) have been taken externally. Complementing the insufficient public support for health and well-being, and/or other societal issues should not be predetermined but left to cultural practitioners to decide. Instead, dictated by the soft power of funding strategic priorities, external funding bodies set the agenda for the cultural sector and determine, via the binding documents of the application briefs, the present and future aspirations and aesthetics of the cultural landscape. The opportunities for institutions and practitioners to "invent" themselves, as Kalmár stated in the interview cited above, or to host the ever-changing and ever-flitting journey of the arts are scarce, if at all existent or possible.[21]

The implementation of the neo-positivist conception of evidence and the ideology of cultural utilitarianism in the cultural sector reveal a systemic distrust in the sector. As we have seen in *The Guardian* incident and in big data projections, evidence and truth-making do not reflect reality and have the prescriptive power to put the future in place based on past events. On the contrary, the cultural sector should be trusted to speculate, withdraw from the restraints of rigid regimes of validation and permissions, entertain chance encounters, unforeseen exchanges, and exposure to the unexpected. During the lockdown, while institutions have largely lain dormant, deposited in a deafening silence, with closed-door exhibitions and sur-rounded by no audience, there has been an opportunity to call for long-overdue trust in the sector. A call to disengage and refuse the subjugating effects of the prescriptive power of funding dependency. To claim non-outcome driven inventiveness for the cultural sector that guarantees an ongoing critical and speculative inquiry into the complex ecology we inhabit and, therefore, affect.

While the future of the arts remains uncertain in a post-Covid-19 society, the legally binding contracts of the funding system and its prescriptive powers should be questioned and contested. Else, when we convene again in the flesh, we may awake to the ultimate take-over of the sector, even more under-staffed and under-resourced, managed by utilitarianism and the prescription apparatus, and asked to, once more, fix the mess.

Notes

1. Luke Harding, "Footage Released of Guardian Editors Destroying Snowden Hard Drives", *The Guardian,* 31 January 2014, https://www.theguardian.com/uk-news/2014/jan/31/footage-released-guardian-editors-snowden-hard-drives-gchq.

2. "Trust, n.", *OED Online,* Oxford University Press, accessed 8 June 2020, https://www.oed.com/view/Entry/207004.

3. Angela Mitropoulos, *Contract & Contagion: From Biopolitics to Oikonomia* (Brooklyn, New York: Minor Compositions, 2012). See also *3 & 4 Will. IV c. 73,* an exhibition of work by Cameron Rowland at the Institute of Contemporary Arts (ICA), London, in 2020; and the work of Brenna Bhandar, Reader in Law and Critical Theory at the School of Oriental and African Studies, University of London, for more on the relationship between the racial materialities of property laws forged through slavery and colonisation. "ICA | Cameron Rowland", Institute of Contemporary Arts, accessed 11 June 2020, https://www.ica.art/exhibitions/cameron-rowland. Brenna Bhandar, *Colonial Lives of Property: Law, Land, and Racial Regimes of Ownership, Global and Insurgent Legalities* (Durham: Duke University Press, 2018).

4. Harding, "Footage Released of Guardian Editors".

5. George Orwell, *Nineteen Eighty-Four* (London: Penguin, 2008).

6. Jean Baudrillard, *The Gulf War Did Not Take Place* (Sydney: Power Publications, 2009).

7. Ariella Azoulay, *Potential History: Unlearning Imperialism* (London: Verso Books, 2019).

8. Michel Foucault, *On the Government of the Living: Lectures at the Collège de France, 1979–1980,* ed. Arnold I. Davidson, trans. Graham Burchell (London: Palgrave Macmillan, 2016).

9. The UK lockdown was announced by the Prime Minister, Boris Johnson, 23 March 2020 via a public address broadcasted on television and online media.

10. "COVID-19 Map", Johns Hopkins Coronavirus Resource Center, accessed 1 June 2020, https://coronavirus.jhu.edu/map.

11. Billy Palmer, "Risky Numbers: The National Reporting of Covid-19", Nuffield Trust, 14 April 2020, https://www.nuffieldtrust.org.uk/news-item/risky-numbers-the-national-reporting-of-covid-19.

12. Tiago Marques, "Por que razão estamos sempre errados em relação ao futuro", *PÚBLICO,* accessed 21 May 2020, https://www.publico.pt/2020/04/22/ciencia/ensaio/razao-errados-relacao-futuro-1913190. Author's translation from the original Portuguese text: "Não podemos acreditar nos números como uma verdade absoluta, porque naturalmente estes correspondem a observações, que podem diferir da realidade, que gostaríamos de observar sem erro, mas que infelizmente está, quase sempre, inacessível".

13. Martina Tazzioli and Daniele Lorenzini, "Critique without Ontology: Genealogy, Collective Subjects and the Deadlocks of Evidence", *Radical Philosophy,* accessed 26 May 2020, https://www.radicalphilosophy.com/article/critique-without-ontology.

14. "Stefan Kalmár, Director of the ICA interviewed by Andrew Durbin, editor of frieze magazine", *Institute of Contemporary Arts* (blog), 1 May 2020, https://www.ica.art/ica-daily/friday-1-may-2020.

15. Maria Lind, Olav Velthuis, and Stefano Baia Curioni, eds., *Contemporary Art and Its Commercial Markets: A Report on Current Conditions and Future Scenarios* (Berlin: Sternberg Press, 2012).

16. Wendy Hui Kyong Chun, "Net-Munity, or The Space between Us... Will Open the Future", *In the Moment* (blog), 20 May 2020, https://critinq.wordpress.com/2020/05/20/net-munity-or-the-space-between-us-will-open-the-future.

17. Carolina Rito, "What Is the Curatorial Doing?" in *Institution as Praxis — New Curatorial Directions for Collaborative Research,* eds. Bill Balaskas and Carolina Rito (Berlin: Sternberg Press, 2020), 44–61.

18. Alan Walker and Steve Corbett, "The 'Big Society', Neoliberalism and the Rediscovery of the 'Social' in Britain", *SPERI,* 8 March 2013, http://speri.dept.shef.ac.uk/2013/03/08/big-society-neoliberalism-rediscovery-social-britain/.

19. Tazzioli and Lorenzini, "Critique without Ontology".

20. A good example of this is the Portuguese Ministry of Culture's response to the Covid-19 pandemic. During the lockdown, commercial and non-commercial sectors were offered immediate financial support. However, the cultural sector was the only one that, instead of having a pack of measures to support its practitioners and institutions, was asked to apply to an open call to compete against each other, with project proposals for new and "innovative, creative and thought-provoking" responses to the crisis.

21. Kalmár, 2020.

References

Azoulay, Ariella. *Potential History: Unlearning Imperialism.* London: Verso Books, 2019.

Baudrillard, Jean. *The Gulf War Did Not Take Place.* Sydney: Power Publications, 2009.

Bhandar, Brenna. *Colonial Lives of Property: Law, Land, and Racial Regimes of Ownership, Global and Insurgent Legalities.* Durham: Duke University Press, 2018.

Chun, Wendy Hui Kyong. "Net-Munity, or The Space between Us... Will Open the Future". *In the Moment,* 20 May 2020. https://critinq.wordpress.com/2020/05/20/net-munity-or-the-space-between-us-will-open-the-future.

Foucault, Michel. *On the Government of the Living: Lectures at the Collège de France, 1979–1980.* Edited by Arnold I. Davidson. Translated by Graham Burchell. London: Palgrave Macmillan, 2016.

Harding, Luke. "Footage Released of Guardian Editors Destroying Snowden Hard Drives". *The Guardian* (31 January 2014). https://www.theguardian.com/uk-news/2014/jan/31/footage-released-guardian-editors-snowden-hard-drives-gchq.

ICA. "ICA | Cameron Rowland". Accessed 11 June 2020. https://www.ica.art/exhibitions/cameron-rowland.

ICA. "Stefan Kalmár, Director of the ICA interviewed by Andrew Durbin, editor of frieze magazine". 1 May 2020. https://www.ica.art/ica-daily/friday-1-may-2020.

Johns Hopkins Coronavirus Resource Center. "COVID-19 Map". Accessed 1 June 2020. https://coronavirus.jhu.edu/map.

Lind, Maria, Velthuis, Olav and Baia Curioni, Stefano, eds. *Contemporary Art and Its Commercial Markets: A Report on Current Conditions and Future Scenarios.* Berlin: Sternberg Press, 2012.

Marques, Tiago. "Por que razão estamos sempre errados em relação ao future". *PÚBLICO.* Accessed 21 May 2020. https://www.publico.pt/2020/04/22/ciencia/ensaio/razao-errados-relacao-futuro-1913190.

Mitropoulos, Angela. *Contract & Contagion: From Biopolitics to Oikonomia.* Brooklyn, New York: Minor Compositions, 2012.

OED Online. *Oxford University Press.* Accessed 8 June 2020. https://www.oed.com/view/Entry/207004.

Orwell, George. *Nineteen Eighty-Four.* London: Penguin, 2008.

Palmer, Billy. "Risky Numbers: The National Reporting of Covid-19". *Nuffield Trust,* 14 April 2020. https://www.nuffieldtrust.org.uk/news-item/risky-numbers-the-national-reporting-of-covid-19.

Rito, Carolina. "What Is the Curatorial Doing?" In *Institution as Praxis — New Curatorial Directions for Collaborative Research,* edited by Bill Balaskas and Carolina Rito, 44–61. Berlin: Sternberg Press, 2020.

Tazzioli, Martina and Lorenzini, Daniele. "Critique without Ontology: Genealogy, Collective Subjects and the Deadlocks of Evidence". *Radical Philosophy.* Accessed 26 May 2020. https://www.radicalphilosophy.com/article/critique-without-ontology.

Walker, Alan and Corbett, Steve. "The 'Big Society', Neoliberalism and the Rediscovery of the 'Social' in Britain". *SPERI,* 8 March 2013. http://speri.dept.shef.ac.uk/2013/03/08/big-society-neoliberalism-rediscovery-social-britain.

Artists and Journalists of the World, Unite! A Kitchen Table Polemic

Ferry Biedermann and Nat Muller

Imagine a curator and a journalist living together and, each morning at the kitchen table, tackling the news of the day from their particular perspectives. Such has been our reality for the past decade or so, and we keep circling back to one particular subject: "fake news", if we should even use that term. Having spent a lot of time together, including on our respective assignments, we have gained some understanding of both sides of the journo-artistic divide, and there certainly seems to be one. However distasteful or inappropriate the term fake news may be, it does touch on many of the core issues that now lay waste to both our fields and to democratic society as a whole. These can, we think, be summarised simply in this old-fashioned way: it is now widely accepted that the ends justify the means, in any aspect of human endeavour, whether ideological, personal, financial, professional, or other. It is now more important to win, or to gain advancement, or to get commissions than to be right. Hence the indifference among large swathes of the public to "alternative facts". Longstanding mechanisms and ways of keeping in check this natural human tendency have been undermined or overthrown in the heedless pursuit of gratification. Art and journalism should both play central roles in countering this tendency but are too often set against each other and thus further marginalised from what they should be doing: providing a counterweight to the manipulation of societies by powerful and other self-interested forces.

Ferry

We accept that cultural trends keep returning in an almost predictable cycle, like low-riding jeans or floral-print dresses. In much the same way, journalism cycles through preferences for a more factual truth and a more emotional truth. Muckraking versus high-brow contemplation, straightforward reporting versus new journalism, contextualisation versus anecdotal storytelling, and so forth. And just when we think we have finally laid the whole objectivity debate to rest, fake news comes and bites us in the arse.

Bear with us, hard put-upon journalists, for a moment, the last couple of decades have not been kind to us and our trade. Falling circulations and viewing figures, causing declining advertising income, have laid waste to the industry, decimating titles in some countries and leading to drastic cutbacks in editorial resources and staff. At the same time, and partly causing the former, the rise of the Internet and social media have contributed to an undermining of journalism as the only source of information on current affairs and a concurrent loss of authority. Weakened news organisations in search of more revenue and an audience have been casting about for a connection with "the people", while at the same time fending off escalating attacks on their integrity and relevance.

For individual journalists, the results have often been dramatic: forced into freelance or part-time jobs, where once the bond between journalist and employer stood for mutual trust and reliability. They are now often reduced to financing their own research and — even — security in conflict zones, where there used to be organisational accountability. Many have had to leave the profession altogether, among others reinforcing the ever-growing army of marketers and media trainers.

A vast informal, unaccountable, and often mercenary network of commentators, spin-doctors and troll factories has sprung up to fill the vacuum left by the retreat of journalism, creating the perfect conditions for fake news to flourish. Actually, I abhor the term fake news, because it lends an air of respectability to good old-fashioned lies and propaganda by affixing the word news. The news, whether on radio or TV, was for decades a moment when large swathes of society came together to partake of the, relatively, unified information stream that created a joint reality. The term "news" had authority, and so its appropriation by those who would rather undermine any common understanding of shared truths is heinous.

At the same time, let's not idealise the journalistic landscape of most of the last century; a substantial number of media organisations have always been more interested in whipping up passions and scaremongering in order to boost their audiences than in responsible news information. And even many of the hallowed news organisations from, let's say, the post-Watergate, golden age of journalism, were blind and deaf to many concerns from minorities, faraway foreign places and the political fringes, to name but a few. Even so, I would argue that more was possible and was actually done than many have given "the mainstream media" credit for.

It's quite astonishing to hear people state that "the media" have never paid attention to this or that. Usually it means that the person stating this has either never bothered to see what was written about it or has a particular agenda to make it seem so, and, sadly, artists sometimes fall into one of those categories and use it as a preamble to their work. Agreed, some important issues went under-reported, and, of course, there are always biases when people are involved, but it was rarely as wholesale as some would like to imply. And where it was, it often was a sign of the times.

While artists are in the avant-garde and are often harbingers of change, journalists are mostly reactive. Even someone like Hunter S. Thompson worked with what he found while on assignments and set out to describe a reality, albeit often using fictional elements; his practice may be as close as a journalist can come to that of an artist.[1] But it is important to remember that writers like Thompson, Tom Wolfe,[2] and others from the New Journalism school still worked within the organised framework of media organisations.[3] They had editors, were given assignments, had fact-checkers (who complained endlessly about Thompson), and they, and the structure they worked in, could be held accountable. There were mechanisms of scrutiny and accountability, however fallible.

This notion becomes less clear when we talk about documentary makers, and slips into oblivion altogether when considering artists who engage with actual events. Nor should it play a role: while art can convey a certain truth, and is valuable for that, it should never be scrutinised or held accountable for presenting The Truth, singular and absolute. Yet, that is exactly what some contemporary conceptual artists seemingly aim to do. From Middle East conflicts to the experience of migrants in Europe to environmental questions, artists have challenged journalism rather than strengthened it. The idea that journalism equals establishment equals vested interests, cover-ups, bias towards the rich and powerful, etc., has firmly been adopted by a segment of the arts and culture community. Interestingly, this discourse is not so far removed from much right-wing criticism of journalism; that it is slanted towards a liberal elite, blind towards the concerns of common people, etc.

This attitude on the part of the, often left-leaning, arts community is counter-productive and unnecessary. A significant amount — if not most — of the journalism produced by the mainstream media is still aimed at uncovering wrongs, speaking truth to power, and holding the powerful to account. Artists and other cultural workers who are interested, are best-served by aligning themselves with this kind of journalism; not by pretending that it does not exist in order to promote their own work. Many artists and writers have been inspired by current events and they have used them as the basis for their work, often taking journalistic accounts as their starting point. A journalist's work only goes so far, and often moves on after the job's done. Artists and fiction writers can offer a much more intimate, emotional, and speculative way into a story. They are not bound by fact, balance, or indeed any kind of professional, and one might even argue ethical, guideline, meaning that they can create worlds and situations that are similar to ours, but not the same — a freedom that would usually place journalists outside the bounds of their profession.

There need not be any opposition between these two approaches. Each is valuable in its own right and can borrow from the other. And of course, they do overlap. But that's where another part of the fake news narrative comes in: activism as one of the longstanding quicksands of journalism. Where does reporting stop and activism begin? For those people on one side of the question, even writing a story about it will smack of activism, while for the activists, the journalist is guilty of betrayal by merely reporting on the issue without unambiguously adopting their stance on it. This can range from human rights to the environment, to consumer issues, etc. With the rise of artivism, or artists who champion political or socio-economic causes in their art, art has entered this contested realm. In their activism, they can be quick to accuse anyone not in alignment with their opposite agenda, hence the link with fake news, of which journalists then stand accused. This is ironic, since there are quite a few artists who adopt fake news scenarios as a tool of their activism. While their goals may be laudable, this also creates more doubt in the public's mind over what is real and what is not.

Art and cultural expression have become ever more instrumentalised in Western society. Governments, sponsors, patrons, and others increasingly want art to affect something in society, or commerce. This also invites pressures to engage in commentary on current issues. But where journalistic institutions and media organisations have, over time, developed mechanisms to distinguish between funding and the editorial, albeit to greater and lesser degrees, the art world has not — at least not in a systematic way. There is no real equivalent in the art world to editorial independence. Curatorial independence may be important to some practitioners but is not in any way codified and is highly dependent on individual preference. Even then, yesterday's critic can be today's curator of a major show at the same institution. This is to say that the art world and art institutions are particularly badly placed to address the challenge of fake news. Where they choose to do so, it can be done in alliance with journalists, but surely not in opposition to them. Any artist or art institution that chooses to tackle the fraught issue of fake news had better make sure that they don't engage in the same practices that they intend to examine. The way out of the fake news conundrum, it has been suggested, is through education, hoping for a more sophisticated and discerning public. If that's so, the last thing we'd want to do, is send mixed messages.

Nat

Many a curatorial text or exhibition press release starts with "the (mass) media depicts A as B", or "the media represents X as Y". This accusatory shorthand is problematic on a variety of levels. Not only does it — for a sector that prides itself on nuance — blatantly conflate all media into one homogenous blob, in which there is no difference between independent, progressive media, and, for example, right-wing and propagandist nationalist outlets. It also does two other things: Firstly, it suggests that art offers a corrective to the blind eyes of "the media", therefore placing artistic production as the harbinger of truth that the media fails to provide. Secondly, and perhaps more troubling, is that it disempowers the Fourth Estate in its mission to inform the public and speak truth to power. It seems that some of the art world claims this role for itself, thereby wittingly, or unwittingly, perpetuating the idea rehearsed by the likes of Trump, Bolsonaro, and company that the media is not to be trusted and is the enemy of the people. While critique of how media organisations operate is healthy and necessary, I echo many of Ferry's concerns and worry how certain segments of the contemporary art world vilify "the media".

My discomfort, however, has more to do with how "veracity" has become simultaneously an obsession, but also a confusion, in artistic practices and curatorial endeavours alike. Art does not necessarily convey truth, other than the truth inherent to its ontic and symbolic representation. What I mean is that art has that wonderful and whimsical quality of being ambiguous, multi-layered, opaque, and falling between a multitude of interpretative cracks. It can speak many truths, or many lies for that matter. In other words, art works in the shadows of meaning. Journalism should, under no circumstances, be afforded this murkiness. But things do become uneasy when art is called on to do just that: convey singular, often didactic, observational realities, and, following on from that, equally singular truths. This is not to say that artists do not provide and enrichen complex points of view on geopolitical, social, and other realities. In fact, most of the artists I have worked with throughout my professional life do so. Their practice seldom strives to substitute journalism; rather, it is about offering something else entirely. Something that should, in fact, never be equated to a 700-word article in a newspaper, or a three-minute item on the news, simply because it plays according to very different rules.

For a start, traditional news cycles and artistic practice have a very different temporal pace. The former is immediate and needs to convey information as soon as possible. It relies on a network of correspondents, journalists, fixers, editors, and media corporations. Analysis happens on the go and often in situ. Artists, in general, need more time to produce work that reflects on a specific situation. Take, for example, a region Ferry and I have worked in for a long time now — the Middle East.[4] It took post-war Lebanese artists about a decade after the Ta'if Accord in 1989, which put an end to the violence of the Lebanese Civil War (1975–1990), to start making work about their experience. Here, their concern was not to provide a final and truthful account of what had happened during the war, but rather to interrogate and unravel the mechanisms of how political and historical narratives are told, or — in the Lebanese context — untold. As visual anthropologist Mark R. Westmoreland notes:

> *Many contemporary Lebanese artists and filmmakers subversively engage visual media in an effort to disrupt the expectations of official and objective "truth telling".*

> *This body of experimental media provides a critical historiography of Lebanon's recent past, particularly in regards to the country's fifteen-year civil war. The intent is not to replace one "false" history with another "true" one, but to go against the grain of sanctioned forgetfulness, commonly referred to as "official amnesia". [...] In a manner of speaking, this particular constellation of artists has kidnapped the historical record in an act of urgent sabotage. This provides a distinctly different approach to the spectacular and sensational reporting provided by Western media.*[5]

While I feel Westmoreland does a disservice to the many (Western) journalists who often risk their lives to get a story out, there are a few things to unpack here. First of all, he offers an insightful differentiation between what artists *do* and what "the media" *does*. And while one can complement or critique the other, they are decidedly not the same. Furthermore, the spectacular and sensational reporting he refers to is, indeed, what unsurprisingly makes the headlines and is what sells. Human-interest stories and long reads just do not feature that prominently, at least not on the frontpages, but this does not mean that they are not out there. There is, however, little difference in how "the media" treats these kinds of headlines and "what sells", and how Western art institutions do. In 2011, during the Arab uprisings, many Western art organisations were just as enamoured by the sexiness and telegenic properties of the revolution as the mainstream media. Budgets were freed, works were commissioned, and many artists found themselves in a situation in which they had to react with immediacy — artistically — to an event that was unfolding. This not only muddled the unique political agency they had claimed there and then demonstrating and camping out, as citizens with political and civic demands, on Tahrir Square in Cairo. But also relegated that very presence to the realm of the representational and symbolic, rather than seeing it for what it was: raw public dissent. An artist joining a demonstration does not automatically turn their protest into a performance piece! Needless to say, much of the work produced in the wake of 2011, excluding all the documentary reports and street art, fell rather flat.[6]

I put no blame on artists wanting to make a living and seizing opportunities in what are precarious circumstances. I do fault institutions and an art world that thinks too much and too little about what art can do. An art world that jumps hastily on cliché bandwagons, whether in condemnation of "the media", or — paradoxically — in the perpetuation of media stereotypes. This haste is often in the service of neoliberal cultural policies, in which artists are required to fulfil all kind of roles: from care-takers, social workers, to now journalists. We would do well to pause and rethink how artists are increasingly institutionally and structurally robbed of the imaginaries and possibilities that are one of the few prerogatives of art. Namely, that art can dance between fact and fiction, and that its truth value lies in a realm other than that of journalism.

Notes

1. Hunter Stockton Thompson (18 July 1937–February 20, 2005), American journalist and author, founder of a sub-genre of New Journalism, called "Gonzo Journalism". It abandoned claims of objectivity and relied on the lived experience of the journalist, who became a protagonist.

2. Thomas Kennerly Wolfe Jr. (2 March 1930–14 May 2018), American author and journalist, one of the main proponents of New Journalism, which he codified in his 1973 anthology, *The New Journalism,* which includes works by, among others, Hunter S. Thompson, Joan Didion, Norman Mailer, and Truman Capote.

3. New Journalism was a genre of journalism that had its heyday in the 1960s and 1970s, taking a more literary personal and subjective approach to reporting. Stories, typically appearing in magazines rather than newspapers, were intensely reported but still hewed to conventional journalism in that the reporter was present in the story, but did not become part of the subject, as they did in Gonzo journalism. Among the criticisms of New Journalism was that it verged on activism.

4. I am aware of the scholarly debates concerning the colonial bias in the term "Middle East". Some scholars, and increasingly more cultural practitioners, opt for West Asia, Southwest Asia, or have replaced the acronym MENA (Middle East and North Africa) with SWANA (Southwest Asia and North Africa) as geographically more neutral terms. Here I use Middle East reluctantly, for the sake of clarity only.

5. Mark R. Westmoreland, "Catastrophic Subjectivity: Representing Lebanon's Undead", *ALIF: Journal of Comparative Poetics* 30, No. 30 (2010): 176–77, https://doi.org/10.2307/27929852.

6. See for a longer discussion Nat Muller, "Stuck in the Middle: How to Review Art Dubai Month 2012", *ARTMargins 2, No. 1* (2013): 106–19.

References

Muller, Nat. "Stuck in the Middle: How to Review Art Dubai Month 2012". *ARTMargins* 2, No. 1 (2013): 106–19.

Westmoreland, Mark R. "Catastrophic Subjectivity: Representing Lebanon's Undead". *ALIF: Journal of Comparative Poetics* 30, No. 30 (2010): 176–77. https://doi.org/10.2307/27929852.

Anarchy near the UK

Bill Balaskas

In January 2016, I was invited by curator David G. Torres and the director of Museu d'Art Contemporani de Barcelona (MACBA), Ferran Barenblit, to create a new work for the touring exhibition *PUNK: Its Traces in Contemporary Art.*[1] I had already participated in the first two iterations of the exhibition in Madrid (Centro de Arte Dos de Mayo) and Vitoria-Gasteiz (Artium Museum), but this time David and Ferran wanted me to exhibit a brand new work, which would address as directly as possible the ideological and aesthetic currency of Punk.[2]

Anarchy near the UK (2016) was the product of this endeavour. In the work, I replicated one of the basic strategies of the Situationists and, later on, the Punk movement: *détournement* — the subversion of propaganda, or turning capitalist media methodologies against the very system that gave birth to them. In the work, this materialises as a spatial counter-collage, where all the news stories on the front page of *The Sun* newspaper of 25 January 2016 have been cut out, leaving intact only the dramatic headline and its reference to anarchy: "Anarchy near the UK". In doing so, it seemed as if the slogans that the Sex Pistols ("Anarchy in the UK") lifted from the anarchist group King Mob, had finally hit the UK. Yet, this was not what had "finally" happened. There was something else that was about to hit the UK — the event that is often evoked as the first major political manifestation of the so-called "post-truth era": Brexit. The title referred to the Calais immigrant "jungle" in France, and to the "refugee crisis" across the Mediterranean. These news stories were framed by the paper as an imminent threat to the UK stemming from the continent — a key debate in the context of the EU referendum campaign, for several months.[3] Notably, in its lead story, *The Sun* accused British anarchists for being "behind a riot that led to 50 migrants storming a cross-Channel ferry in Calais".

The amended front page of *The Sun* was accompanied by a display vitrine, which featured a series of objects and memorabilia alluding to the stories that had been removed from the newspaper cover. These included the news about a famous footballer (Wayne Rooney) becoming a dad to a baby boy; a major football club (Manchester United) being in crisis due to bad results; a famous singer (Harry Styles of the boyband One Direction) dating a fashion model and reality television celebrity (Kendall Jenner of the Kardashians); a story about Google's tax avoidance in the UK; and a story about an elderly Lotto winner.

By juxtaposing the dramatic title of the tabloid newspaper with material representations of the removed stories, *Anarchy near the UK* aimed at highlighting the absurdity of today's world, in which spectacle has thoroughly replaced facts. As the audience is called to "reconnect" the missing news stories and infer their meaning, the work exposes the challenges that relate to not only unveiling the truth, but also communicating it — making it public. In this way, the work also highlights the responsibility of the viewer-audience, as a key agent in this process; an invitation to active citizenship and criticality, inspired by Punk's spirit of anti-authoritarianism.

Anarchy near the UK was exhibited for the first time six weeks before the EU referendum in the UK and six months before the Presidential election in the US. Notably, these were the two events cited as catalysts by the Oxford Dictionaries in their declaration of "post-truth" as Word of the Year 2016, just a few days after the election of Donald Trump as the 45th President of the United States.[4] Despite primarily constituting a response to the ideological antagonisms during the Brexit campaign, the investigation of media spectacle and populism in the work stems from conditions that both precede and go beyond the aforementioned political events. Contemplating such conditions requires "pausing" the temporalities of our increasingly social-media-driven communication, in order to look *at* and look *into* at the same time. (Re-)discovering what lies in front of us, and what has been omitted, is the key challenge and duty that we have to contend with in this contradictory "post-truth era".

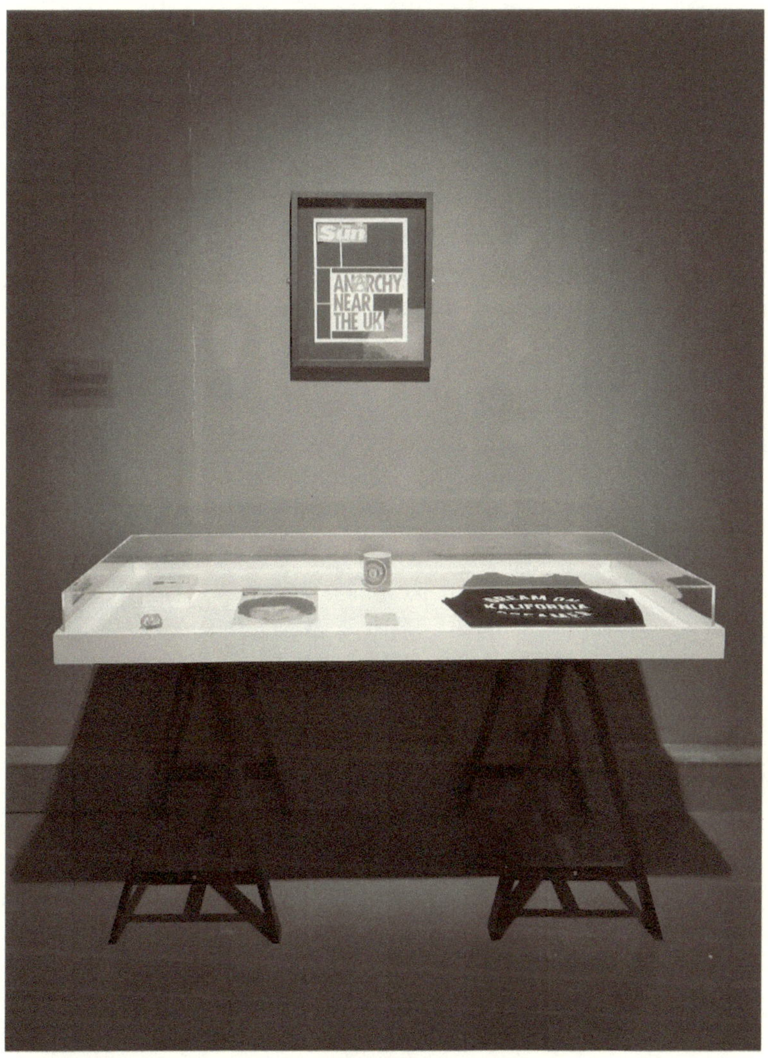

Figure 1. Bill Balaskas — *Anarchy near the UK,* 2016, mixed-media installation. Frame: amended front page of *The Sun* newspaper (25 January 2016). Display case: Manchester United mug; Lotto ticket; baby-boy pacifier; mask of One Direction singer Harry Styles; a copy of *Are You Smart Enough to Work at Google?* (2012); and a t-shirt designed by celebrity personality and model Kendall Jenner.

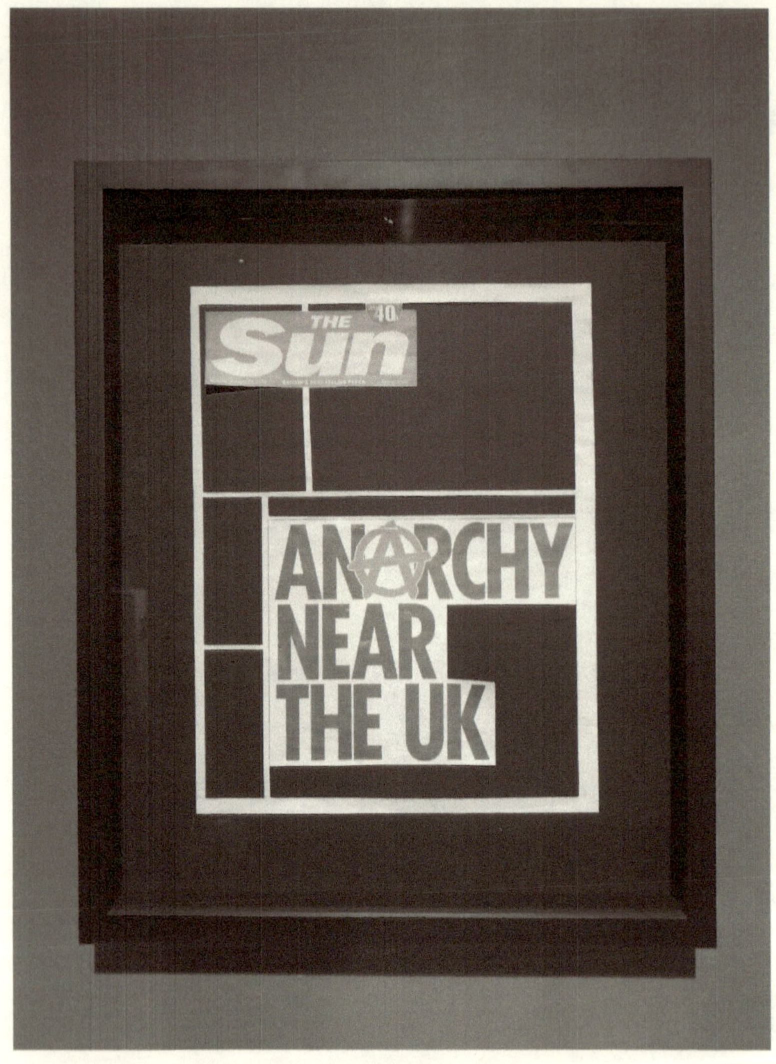

Figure 2. Bill Balaskas — *Anarchy near the UK,* 2016, mixed-media installation (detail). Frame: amended front page of *The Sun* newspaper (25 January 2016).

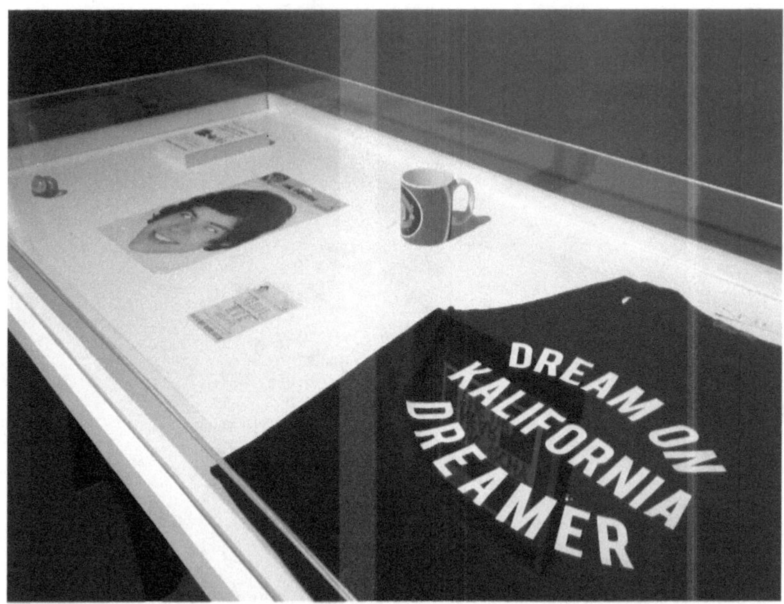

Figure 3. Bill Balaskas — *Anarchy near the UK,* 2016, mixed-media installation (detail). Display case: Manchester United mug; Lotto ticket; baby-boy pacifier; mask of One Direction singer Harry Styles; a copy of *Are You Smart Enough to Work at Google?* (2012); and a t-shirt designed by celebrity personality and model Kendall Jenner.

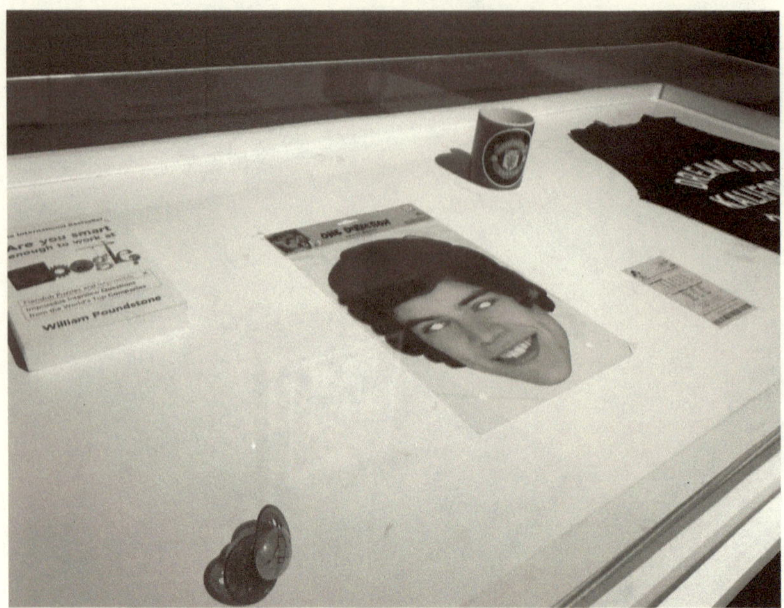

Figure 4. Bill Balaskas — *Anarchy near the UK,* 2016, mixed-media installation (detail). Display case: Manchester United mug; Lotto ticket; baby-boy pacifier; mask of One Direction singer Harry Styles; a copy of *Are You Smart Enough to Work at Google?* (2012); and a t-shirt designed by celebrity personality and model Kendall Jenner.

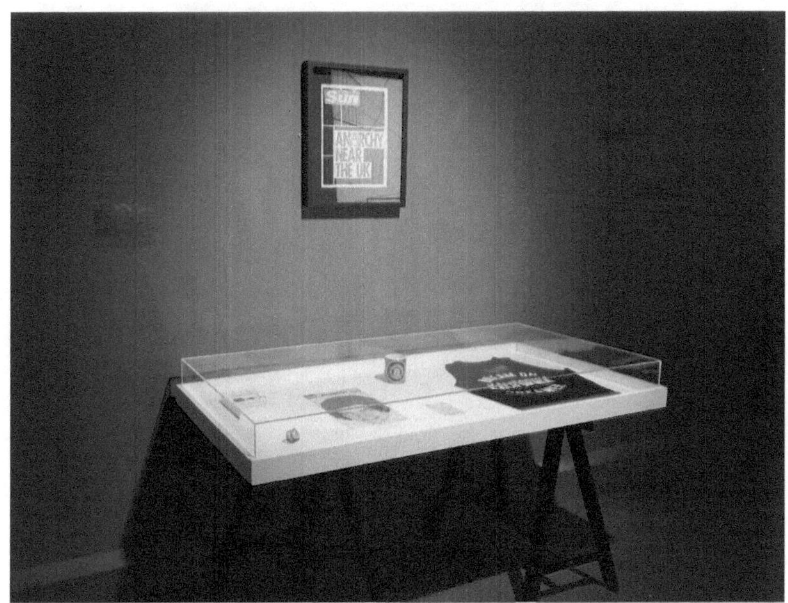

Figure 5. Bill Balaskas — *Anarchy near the UK,* 2016, mixed-media installation. Frame: amended front page of *The Sun* newspaper (25 January 2016). Display case: Manchester United mug; Lotto ticket; baby-boy pacifier; mask of One Direction singer Harry Styles; a copy of *Are You Smart Enough to Work at Google?* (2012); and a t-shirt designed by celebrity personality and model Kendall Jenner.

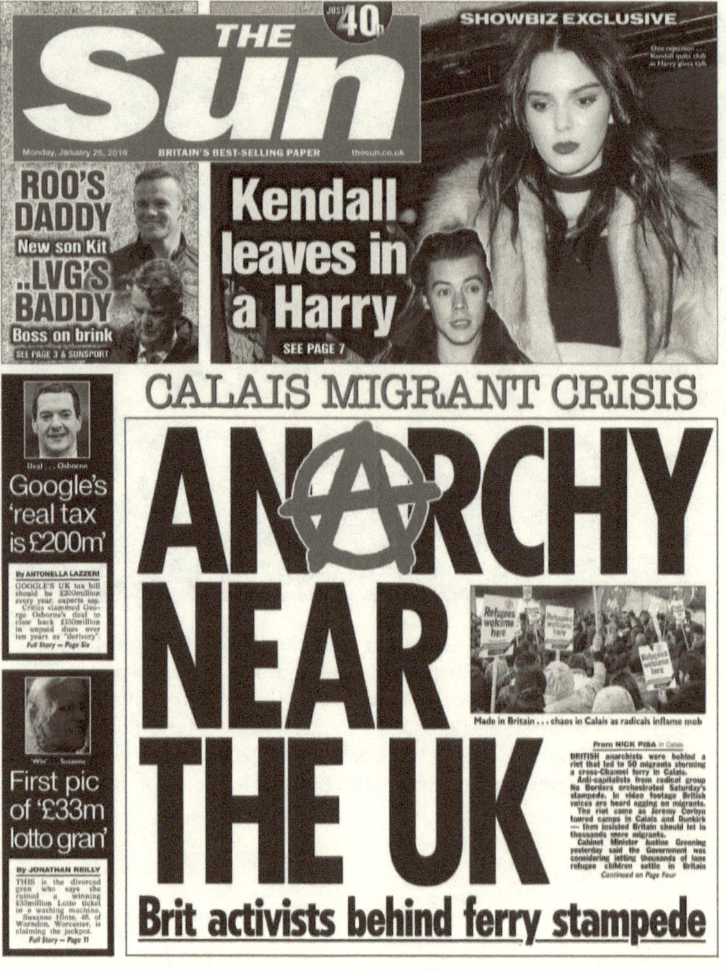

THE Sun

JUST 40p

Monday, January 25, 2016 BRITAIN'S BEST-SELLING PAPER thesun.co.uk

SHOWBIZ EXCLUSIVE

One rejection . . . Kendall quits club as Harry gives talk

ROO'S DADDY

New son Kit

..LVG'S BADDY

Boss on brink

SEE PAGE 3 & SUNSPORT

Kendall leaves in a Harry

SEE PAGE 7

Deal . . . Osborne

Google's 'real tax is £200m'

By ANTONELLA LAZZERI

GOOGLE'S UK tax bill should be £200million every year, experts say.

Critics slammed George Osborne's deal to claw back £130million in unpaid dues over ten years as "derisory".

Full Story — Page Six

'Win' . . . Susanne

First pic of '£33m lotto gran'

By JONATHAN REILLY

THIS is the divorced gran who says she ruined a winning £33million Lotto ticket in a washing machine.

Susanne Hinte, 48, of Warndon, Worcester, is claiming the jackpot.

Full Story — Page 11

CALAIS MIGRANT CRISIS

ANARCHY NEAR THE UK

Made in Britain . . . chaos in Calais as radicals inflame mob

From NICK PISA in Calais

BRITISH anarchists were behind a riot that led to 50 migrants storming a cross-Channel ferry in Calais.

Anti-capitalists from radical group No Borders orchestrated Saturday's stampede. In video footage British voices are heard egging on migrants.

The riot came as Jeremy Corbyn toured camps in Calais and Dunkirk — then insisted Britain should let in thousands more migrants.

Cabinet Minister Justine Greening yesterday said the Government was considering letting thousands of lone refugee children settle in Britain

Continued on Page Four

Brit activists behind ferry stampede

Figure 6. The original front page of The Sun newspaper on 25 January 2016 that was modified in *Anarchy near the UK* (2016). The front-page lead story reads: "BRITISH anarchists were behind a riot that led to 50 migrants storming a cross-Channel ferry in Calais. Anti-capitalists from radical group No Borders orchestrated Saturday's stampede. In video footage British voices are heard egging on migrants. The riot came as Jeremy Corbyn toured camps in Calais and Dunkirk — then insisted Britain should let in 3,000 thousand more migrants. Cabinet Minister Justine Greening yesterday said the Government was considering letting thousands of lone refugee children settle in Britain [...]".

Notes

1. The exhibition *PUNK: Its Traces in Contemporary Art* took place at Museu d'Art Contemporani de Barcelona (MACBA) from 13 May–25 September 2016 (http://ca2m.org/en/archive/item/2288-punk-sus-rastros-en-la-creacion-contemporanea). Its first two iterations took place at Centro de Arte Dos de Mayo (CA2M), Madrid, from 25 March–4 October 2015; and at Artium Basque Museum-Center of Contemporary Art, Vitoria-Gasteiz, from 23 October 2015–31 January 2016. Link to the exhibition held at ARTIUM, Basque Centre-Museum of Contemporary Art (https://www.artium.eus/en/exhibitions/item/55870-punk-sus-rastros-en-el-arte-contemporaneo).

2. In the first two iterations of the exhibition, Bill Balaskas exhibited his installation work *ÉCANOMIE,* 2011, mixed-media, site-specific installation: paint, brushes, plastic buckets, wooden plinth, dimensions variable.

3. In one of the most famous anti-migration incidents in the build-up to the EU referendum, the UK Independence Party (UKIP) published a poster showing a queue of predominantly non-white migrants and refugees with the slogan: "Breaking point: the EU has failed us all". The poster was created in the context of a debate that regularly focused on Turkey, and the prospect of the country joining the EU in the future. Notably, an official poster about this issue was published by the Leave campaign on Monday 23 May 2016. The imminence of Turkey's membership of the EU was, at the time, rejected by EU governments and officials. By the spring of 2021, this prospect appeared even more distant. See: https://www.theguardian.com/politics/2016/jun/16/nigel-farage-defends-ukip-breaking-point-poster-queue-of-migrants; and https://www.ft.com/content/f264be32-2cc6-11e6-bf8d-26294ad519fc.

4. Oxford Languages. "Word of the Year 2016". https://languages.oup.com/word-of-the-year/2016.

References

Crawley, Heaven, and Skleparis, Dimitris. "Refugees, Migrants, Neither, Both: Categorical Fetishism and the Politics of Bounding in Europe's 'migration Crisis'". *Journal of Ethnic and Migration Studies* 44.1 (2017): 48–64.

d'Ancona, Matthew. *Post-Truth: The New War on Truth and How to Fight Back.* London: Ebury Press, 2017.

Davitti, Daria. "Biopolitical Borders and the State of Exception in the European Migration 'Crisis'". *European Journal of International Law* 29.4 (2018): 1173–1196.

Demos, T. J. *The Migrant Image: The Art and Politics of Documentary during Global Crisis.* London: Duke University Press, 2013.

Dines, Nick, Montagna, Nicola, and Vacchelli, Elena. "Beyond Crisis Talk: Interrogating Migration and Crises in Europe". *Sociology* (Oxford) 52.3 (2018): 439–447.

Eberwein, Tobias, and Fengler, Susanne, eds. *Media Accountability in the Era of Post-Truth Politics: European Challenges and Perspectives.* New York: Routledge, 2019.

Farkas, Johan, and Schou, Jannick. *Post-Truth, Fake News and Democracy: Mapping the Politics of Falsehood.* New York: Routledge, 2020.

Greenhill, Kelly M. "Open Arms Behind Barred Doors: Fear, Hypocrisy and Policy Schizophrenia in the European Migration Crisis". *European Law Journal: Review of European Law in Context 22.3* (2016): 317–332.

McIntyre, Lee. *Post-Truth.* Cambridge, MA: The MIT Press, 2018.

Squire, Vicki. *Europe's Migration Crisis: Border Deaths and Human Dignity.* Cambridge: Cambridge University Press, 2020.

Natalie Bookchin in Conversation with Alexandra Juhasz: Performances of Race and White Hegemony on YouTube

Natalie Bookchin and Alexandra Juhasz

In November 2019, in her house in Bedford-Stuyvesant, Brooklyn, I joined friend and fellow artist Natalie Bookchin for a conversation about her installation and film *Now he's out in public and everyone can see.* The installation, which premiered at Los Angeles Contemporary Exhibitions (LACE), a venerable Los Angeles art space, in 2012, was remade into a film and released as a DVD double feature along with her film *Long Story Short* by Icaras Films in 2016. Our loose and lively conversation was recorded and transcribed and forms the basis for what follows. We have been in conversation about digital culture, YouTube, video, social media, art, and politics for many years, and thought that this would be a productive way to gain new insight into Natalie's project and its themes of internet publicness. Her work is especially relevant today given the current landscape of online media and its relationship to our troubling political climate. It is telling that the work we discuss was made in 2012 (and then 2016), and that the work that had cemented our friendship and ongoing professional engagements was made even earlier in social media history — my book *Learning from YouTube* (2011),[1] and Natalie's significant body of YouTube-built video works from the early 2000s. These time shifts, in a quickly changing media landscape mapped by our work alongside it, and our shared, if changing, senses of publicness, possibility, and politics form the heart of a conversation that anticipates the American reckoning on anti-Black racism and violence that was renewed and intensified in summer 2020.

Alexandra Juhasz: We're here to talk about *Now he's out in public and everyone can see.* I'm really delighted. To begin, can you describe *Now he's out in public* for someone who's never seen it?

Natalie Bookchin: In the installation, 18 monitors of different sizes hang at varying heights and distances around the perimeter of a darkened room. Monitors light up as vloggers appear on the screens, standing or sitting in bedrooms, bathrooms, and other domestic spaces. They begin to recount incidents of some concern apparently involving a famous Black man, forming a chorus of voices, faces, and opinions envelope the space. Voices ricochet around the room, producing a rhythmic cadence and an affective sonic and visual environment. The space feels crowded and charged with impassioned, sometimes threatening, and antagonistic chatter. Periodically, the narrators speak in unison, other times one speaker echoes, completes, or contradicts a previous speaker's thoughts, or adds details or comments to a remark. The film version, on the other hand, is mostly composed of extended close-ups, relying on a dense layered soundscape of voices to create a claustrophobic and antagonistic space.

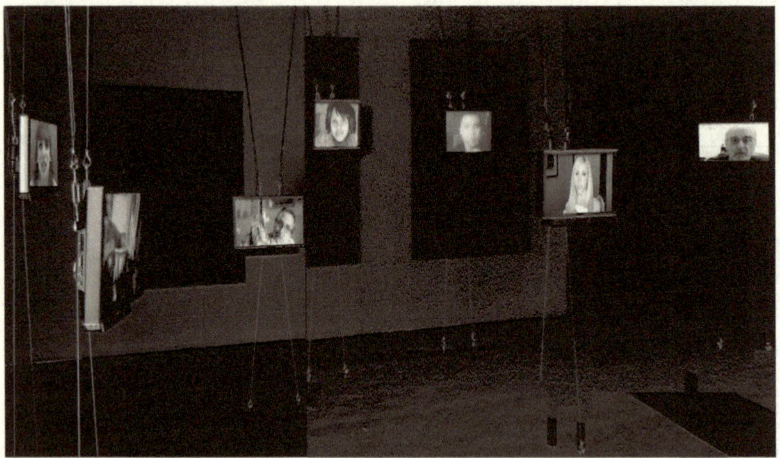

Figure 1. *Now he's out in public and everyone can see,* installation.

AJ: The installation and the single-channel work are both built from hundreds of "narrations" made by everyday YouTubers that originally took the form of vlogs. From these private stories and testimonies do you think it is fair to say that you build a "public narrative"?

NB: Yeah, public and collectively produced. The "story" is a composite of reactions, responses, reenactments, and descriptions of a series of incidents and a racist conspiracy theory (initialed and promoted by our current president) that went viral involving four famous African-American men. I removed the names of the men and edited different commentators together to create a winding narrative about a famous, rich, Black man who, whatever he has done, or is, provokes very strong reactions from a disparate public who can't stop talking about him.

AJ: Albeit a man who keeps slipping between your fingers...

NB: The work also reflects the incredible contagion of media narratives involving race, and how social media revels in them, spreading, circulating, and prolonging their lives. The narrative is, in fact, composed of YouTube narrated stories, lies, rumors, projections onto, and incidents involving four black celebrities in completely different fields — there is a politician, a golf player, an academic and TV celebrity, and a singer. The narrative I build focuses on the srepetition in the language used as vloggers recite and perform their narrations, and the way that those performances diverge along racial and gender lines.

AJ: As well as stylistics and formats taken up online to discuss and share.

NB: Right. Language is arranged in the work around common themes, shared and overlapping rhetoric, words, and phrases, producing a kind of catalogue of popular tropes used to discuss race and Blackness. The speakers debate how well the man is managing his status and position as a leader and role model. For some, he has been treated unfairly, held to impossible standards. Others say he's been a disappointment and hasn't lived up to expectations. Some say he was arrested outside his own home after being mistaken for a burglar by a white neighbour. Others insist that he crashed his car into a hydrant outside his home, at which point his white wife began smashing something — himself? a window? — with his golf clubs. Throughout, the man's identity, especially his status as a Black man, is repeatedly called into question. He is referred to variously as: "a fucking god", "the Messiah", "a black male", "the motherfucker", "a black guy", "not black", "half white", "an African American", "half-African American", "56% white", a "Muslim", "a mask", "a fraud", "more of a white guy", "one of us", "not really one of us", "a usurper", "a socialist", "a paedophile", "a kid at heart", "an idol", "a hero", "a role model", "the second coming", "a wonderful guy", "the negro", "boy", "you", "a human being", "the Black Prince Charles", "a fucking billionaire", and "the most desirable guy — as far as females are concerned — in sthe world".

AJ: These many interpretations are edited into a single composite narrative that unfolds across 18 screens relayed by what feels like uncountable speakers. Why create a composite of four African-American men and their four public scandals, and why don't you name the man?

NB: Weaving the various scandals and rumors together and removing the names suggest that the specifics don't really matter. The man in question is a figment of the speakers' and the viewers' imaginations, composed of rumours and gossip, speculations, and judgements. The language and the stories themselves keep repeating — different man, different incident, but same old story. Just as vloggers try to "authenticate" the man, so do viewers of the installation. But as soon as an audience recognises one story, it changes and the man in question "slips from their fingers". An authentication can never happen; viewers can never "master" the narrative. Just as they can't see all the speakers in the installation at one time — there is always someone speaking out of view, in another corner — they never "get" the whole story. The view is always partial and fragmented.

AJ: As a viewer of *Now he's out in public,* you can't help but note the differences between the famous Black men who are subjects of the media and the ordinary people who are making media about them. Of course, one of the prime motivators of social media in general, and YouTube specifically, has been a vague promise of Internet fame. Each vlogger seems to enact, or anticipate their own possible parallel fame, an elevated state signaled as available to all by a social media still in its infancy, one full of potential and desire and hope. They ridicule, analyse, pick apart, and somehow also hope to be *him,* even though this fame, and his publicness, as your piece suggests, produces his or their downfalls. That said, the piece also depicts the ambivalence, anger, jealousy, and ridicule focused on these men in particular because *Now he's out in public* is less about being famous per se as it is about being famous while Black.

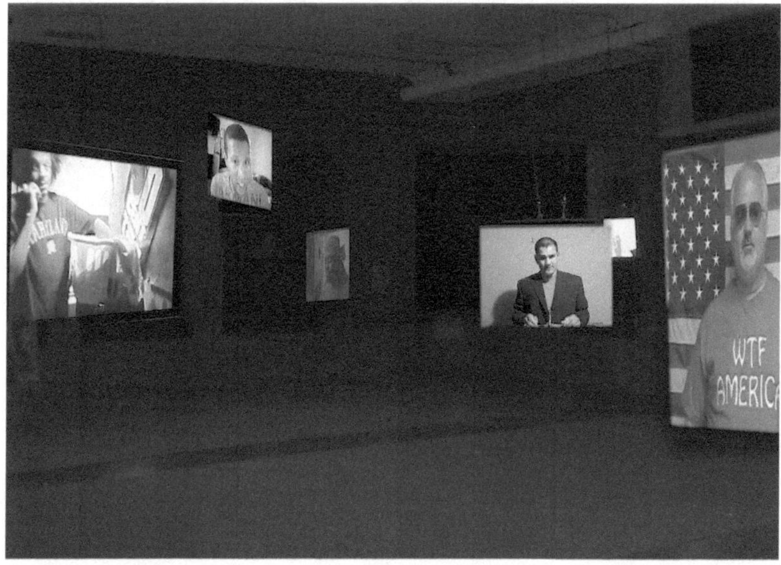

Figure 2. *Now he's out in public and everyone can see,* installation.

NB: Yes, absolutely. The work explores how antagonistic performances of race and white dominance were a significant part of online spaces like YouTube even in its infancy. The piece also looks at how white anger against so-called elites and the wealthy, from the beginning, online, took on a racist tone, and due to their volatility, and thus their tendency to be watched and spread, were promoted and amplified on YouTube. Many of the vlogs in the piece were produced just after Obama's election and the anxiety and discomfort of some of the white vloggers as they discuss Black success is palpable. The Black vloggers, on the other hand, mainly express discomfort that the man has been caught in public in some unnamed act of transgression. They fear for his publicness. What it boils down to in each of these so-called scandals, is that to be a Black man and in public *is* the scandal. Things start to go wrong, as one blogger states, when the man "steps outside his door".

AJ: Your installation builds from a set of interests and practices you had been working on for quite a while: making art out of YouTube videos and vlogs. Can you talk about your earlier work and how you began to develop your now signature method, voices woven like a chorale where you arrange multiple speakers (found in the wild) to say the same word or the same sentiment in unison, or relay, or even opposition, as if they were choreographed or scripted?

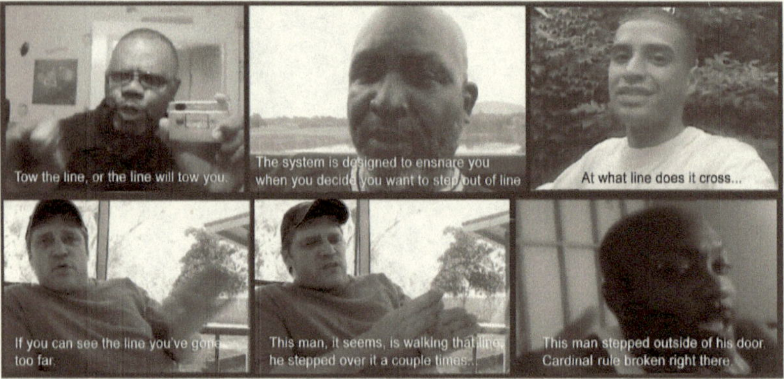

Figure 3. *Now he's out in public and everyone can see,* captioned film stills.

NB: You're referring to work like *Testament and Mass Ornament,* both from 2009 — video installations in which I constructed a chorus and a mass dance (respectively) out of numerous found online videos. On YouTube, ordinary people began making and sharing videos, spontaneously posting their thoughts and opinions, or performing for the camera and to the world. The videos felt like inadvertent, or found self-portraits to me, and suggested a collective yearning for publicness. Yet, these collectivist yearnings were mostly buried beneath interfaces and designs that highlight and reward single users. On YouTube, users have their own "channels", subscribers, and playlists, and are forced into competition with each other for likes and views. I was interested in the tension between these public and collective desires, and the site's design constraints which isolated and monetised single "users". I wanted to depict overlapping subjectivities and interconnectedness — something that was hard to see in viewing single videos alone.

Now he's out in public is an extension of this earlier work, but it also goes in a different direction. The earlier work focused on people intentionally exposing or revealing something about themselves, highlighting precarity, vulnerability, and desire for connection. The vloggers in *Now he's out in public* mostly appear less concerned with connecting with others than with broadcasting their own opinions. Instead of talking about themselves, they are self-appointed judges, or protectors, of others. They mostly seem oblivious about what they are exposing about themselves. When they speak in unison, particularly with ad hominem attacks on the man in question, it can feel less like a chorus and more like a digital mob.

AJ: So true! At times it's hard to be in the room with them. But at the same time, there's a way in which you are providing a service at a point in digital history that the platforms are not yet able to produce for themselves, or for us. You're making connections (by hand!) that happen now, something like ten years later, through algorithms. In 2012, your project as an artist was to find, show, and make into collectivity for and from a place where that was not yet publicly renderable, even as these very platforms were encouraging and then collecting masses of individual voices and data about them under the hood.

NB: Right, although it is not that the so-called platforms weren't able to produce images of collectivity; I'm just not sure they have a financial interest to do so. Where is the revenue stream in that? The term "platform", which companies like YouTube, Facebook, etc. use to describe themselves, suggests a neutral, horizontal base onto which the media we share freely circulates. But we know that is not what happens. Content with more views rises to the top, while less "popular" material is buried. I created montages that attempt to make visible associations that might otherwise not be seen or noticed.

AJ: Associations known and used by the corporations! The bullies. And sometimes movements, I suppose. In that earlier work, you revealed the vlog's intimacy and a connection between that intimacy and the isolation of YouTubers. Your service as an artist was to connect people, ideas, words, themes, feelings. And so, your work reveals a tension between the intimacy of the encounter between people and their cameras, between people and their videos and their imagined audience, as well as the aloneness of these subjects — so much of your work shows a person in their own room mirroring us in ours — and what was still a live belief in a promise of publicness.

NB: I think the willingness with which people exposed themselves in the early days of social media carried with it a hope that the Internet and social media would build community and social relationships that are missing in our society. But we were sold on a lie. Instead of opening the world up, the big tech companies who took over the internet make the world more constrained, narrow, and limited, sequestering each of us into our own micro-targeted universe. That isn't to deny that some progressive communities did form and still thrive despite the tech takeovers. Black Lives Matter, #MeToo, Black Twitter, and many of the progressive protest movements around the globe make use of social media. But even so, right now the racists, propagandists, and nationalists empowered by big tech have been threatening democracy around the globe. It's finally become common knowledge that Silicon Valley won't save us.

AJ: It's strange to see something we've both known and spoken about for so long — in public, in art, in writing — now, finally, being understood as itself a social, or public truth. As the perception of digital and social media has shifted for its everyday users, did your approach to an analysis of it also change? For instance, before *Now he's out in public,* you had been showing your work made from vlogs as projections on walls in galleries or museums. What moved you to build this argument into an installation with multiple screens? Why did you have to spatialise what was changing for people and video on social media? Is this related to what the editors of this book propose as a "traumatic fragmentation of the social body" following the global financial crisis of 2008?

NB: The installation of *Now he's out in public* conjures a mass that is fragmented and dispersed — a reality shattered into shards of opinions. There is no centre, no shared or agreed upon truth. Instead, there are clusters of opinions, instances of partial unity that quickly scatter and break apart. There are instances, for example, when all the speakers on all 18 monitors say the same thing at the same time, but this unity is brief, and quickly replaced by smaller groups of speakers where one group claim one truth, while another claim a different one.

AJ: Can you discuss another aspect of the installation: the embodied experience of the viewer moving through, and interacting in the room with the vloggers, the physical experience of a narrative unfolding in space? Being in the installation felt almost as if you were in the room with each of the speakers.

The viewer became part of this unseemly chorus in a way that hadn't been true with your previous videos, where we watched from the outside looking in.

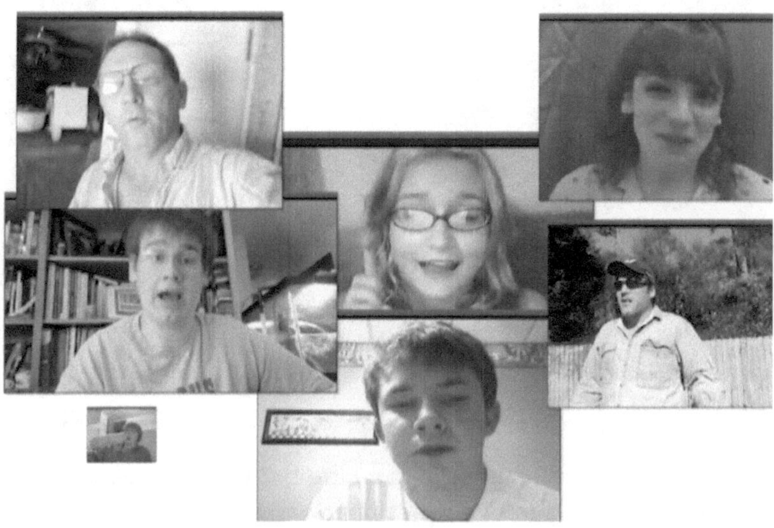

Figure 4. *Now he's out in public and everyone can see,* film still.

NB: Yeah, with this work, instead of multiple frames of videos on a single screen, the montage is spatialised, and viewers must traverse the space to see and experience the work. In this way, the viewers' bodies are activated. This embodiment reiterates the themes of the work, which suggests that bodies, and embodiment, matter. There were a lot of claims in the early days of the Internet that, with experience becoming increasingly virtual, physical bodies no longer mattered. Related, when Obama was first elected as president, many claimed that we were entering a "post-racial" era, one where race, where the historical specificity of bodies no longer mattered. In *Now he's out in public,* bodies are affected. In order to experience the visceral, affective installation, you also have to *be there* in the flesh. The narrative points to racial violence against specific bodies in public space, even virtual public space, suggesting that language has an impact on real bodies, including — especially — Black bodies under scrutiny.

AJ: But things have changed since then. We are now in an age of social media that's fully disembodied. Twitter and Instagram are populated by unseen speakers.

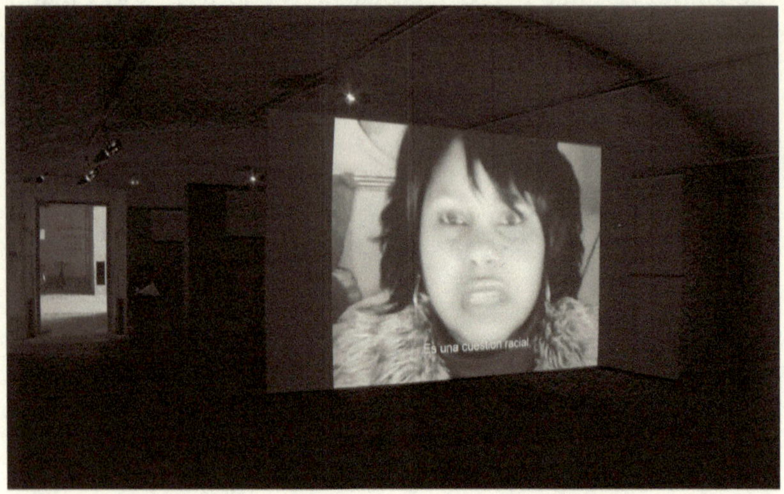

Figure 5. *Now he's out in public and everyone can see,* installation.

NB: Right, on Twitter, people can hide behind handles, and you can never be sure if a tweet's been written by a person or generated by a bot. In *Now he's out in public,* viewers are face-to-face with the speakers. You can look behind them into their homes, and at their things arranged or left in the frame by accident or indifference. I would look for these details as I edited, as well as for moments when the vloggers were silent, when they lingered, hesitating, or sipping on a drink, glancing at themselves on the screen, adjusting props, arranging the camera. I searched for moments when people stopped performing, or when they slipped out of the performance — learned by heart from Fox News or whatever other media they were watching — for moments where they let their guard down, when you can detect instances of uncertainty or vulnerability. On Twitter and Instagram, those moments are much harder to find. It's much easier to hide behind poses and talking points.

AJ: Agreed! In vlogs, we get a chance to see the human being at the end of the chain of signification. In our recent post-truth era, we can't as easily get there: to the person who made and said shit. Now everything's possible to say, but by whom? We need systems that can help us render what just might stay live between two people. Yes, words, and bodies, and places, but also affect. That is one reason why my own work on fake news has turned to poetry and performance over indexical images.[2]

Instagram and Twitter offer certain freedoms, but performative embodiment is not one of them. You register that for us when we embody a room with these people. But later, you decided to make this work into a single-channel piece. Can you talk about what happens thematically when you flatten and make linear our encounter? What do people learn when they engage with these narratives as a film?

NB: I decided to remake *Now he's out in public* as a film the summer before Donald Trump was elected as president. The themes of the work — the fracturing of truth, and the growing prominence of racist speech and angry white crowds, the increased polarisation, misunderstandings, and isolation among our population scaled up thanks to the tools of big tech — seemed increasingly relevant. Even though the work was made before these themes became such a prominent part of the public conversation, I thought it might add something to the current debates. The installation is complex and expensive to install — and impossible to document — so I decided it would be worth making it into something accessible: a film. I released the film in 2017 as a double feature with another film of mine, *Long Story Short,* which I had finished the year before.

AJ: Those two works share an editing language that you refined across this body of work, but they are almost polar opposites in the nature of the speaking and visibility of the voice in video. To make *Now he's out,* you found people who spoke online but remain anonymous to you and us; while for *Long Story Short,* you shot the footage and, thus, the speakers become known to you and then us through a kind of loving, intimate support in your editing that you had not given to the video of, and by humans you had worked with previously.

NB: In *Long Story Short,* I interwove interviews I'd shot with over 100 people about their experiences and perceptions of, and insights into living in poverty. People talked about what they thought the media got wrong in their depictions and what they wanted to see instead. Each interview lasted over an hour. On YouTube, videos used to be limited to 10 minutes or less, and most of the vlogs I collected were a lot shorter. Part of the strangeness of vlogs is that people are alone, talking to themselves, hoping for someone to hear. In *Long Story Short,* I was in the room, so people were, at least at that moment, being heard.

AJ: The affect that is so live in the video you shot for *Long Story Short* seems critical in relation to the loss of place and person that currently defines social media. Looking back at *Now he's out in public,* it anticipates a now commonly understood alienation in the face of social media's promise of community. Does it also anticipate possibility and hope in terms of people's access to democracy via technology and representation?

NB: We all now know and have experienced the significant negative effects of technology controlled by big corporations and repressive governments. I think the hope is in local embodied practices where protest and resistance happen both in media space and in person with other people. I'm thinking, for example, of practices where groups of people find ways to use technology and commercial platforms to reinforce and sustain visibility and already existing connections around particular issues or identities. I'll give you an example from a project I am currently working on. It is a film with the working title *Sonidos Negros (Black Sounds)* that I'm making in collaboration with a Roma association in Spain, Lacho Bají, and a Spanish artist collective, LaFundició. Together, we are developing a collective cinematic portrait of, and with, the local Spanish Roma community, exploring modes of representation of, and by, a group of people long stigmatised and discriminated against by the majority white Spanish society. Although Roma history has for centuries been repressed by the Spanish majority, local Roma groups are actively reconstructing their hidden pasts — their histories and traditions in Catalonia and their deep roots in Spain. People use Facebook and WhatsApp groups to share instances of "antigypsyism" and pro-Roma material. They are not looking to these sites with the goal of creating community that doesn't yet exist, but rather to sustain existing connections. So, these sites are not substitutions for "community", but rather media channels for distributing forms and content that aren't easily seen elsewhere. The film will offer a radical pastiche that utilises visual aesthetics inspired in part by social media feeds. In contrast to stereotypes about "gypsies" as primitive and pre-modern, the film counters mainstream and stereotypical depictions of the Roma as anti-modern and underdeveloped, out of touch with current trends, technologies, and realities. We're also exploring how these tools are appropriated by groups such as the Roma, whose vitally active community life and economies of sharing and giving offer significant lessons for, and radical alternatives to hyper-individualism and dehumanising neo-liberal economic models.

AJ: All of your work allows us to see how places, bodies, and media are critical, if we are to retain a public that can nourish, engage, and empower us. Thank you.

Notes

1. Alexandra Juhasz, *Learning From YouTube* (Cambridge, MA: The MIT Press, 2011), http://vectors.usc.edu/projects/learningfromyoutube/ [born-digital "video-book"].

2. See: fakenews-poetry.org; and Alexandra Juhasz, Ganaelle Langlois and Nishant Shah, *Really Fake!* (Minneapolis: University of Minnesota Press, 2021); and *Open access,* Meson Press, 2021: https://meson.press/books/reallyfake.

Lying as Truth: Cervantes as Co-Author of Don Quijote

Mieke Bal

Authorship multiplied

In a recent article, I made a plea for the revitalisation of the notion of authorship since Roland Barthes and Michel Foucault declared the poor chap dead in the late 1960s. That demise was a bit premature. My motivation and condition for bringing the author to life again: authorship is important, not for legal reasons such as copyright, but to understand its cultural status as multiple. For Barthes, as per his final sentence, the author had to die in order to recognise the role of the reader — the reader as co-author. Rather than a passive public, readership is a contribution to the only existence a literary text can enjoy: as a cultural object that, due to the essential performativity of language, becomes the thing around which the social buzz of readers, reviewers, and also, others who remake the work as film, theatre, or otherwise, is active. With this multiplicity in mind, it is really worth re-reading Barthes's and Foucault's articles again, seriously. They have a lot of important things to say.[1]

If I write this article mainly a apropos of "my own" video installation *Don Quijote: Tristes Figuras (Sad Countenances)* from 2019, it is to explain some of the many tentacles of the issue of multiple authorship and its/their relation to the pub-lics. The installation, which is on exhibition as I am writing this text, is not "my own" any more than the literary text on which it is based — or rather, from which it takes off and to which it responds — is Cervantes's own. The Spanish author gave the world its first novel, and its first international best-seller; and, in my view and as motivation for the video work, the first literary, poetic account of trauma — of course, without using the word, which was not invented until centuries later, but in its form. The world has not finished responding to it in ways that can be helpful for the social fabric within which the novel exists. *Don Quijote,* as I will abbreviate the title, entered a world of violence, war, and slavery, which continues today. The form of the novel addresses and, thereby, creates a public that is much needed, if art is to have any significance at all for the world.

As it happens, the author wrote it *after* five and a half years of slavery he had been subjected to, *after* being captured when the ship cheerfully sailing back to Spain *after* the victory at the battle of Lepanto (1571) was attacked by Corsairs, and only Miguel de Cervantes and his younger brother Rodrigo didn't make it back home. The three times "after" in the preceding sentence indicate a temporality and a causality that is also multiple. He could not have written the novel as we have it without that horrific experience. And in case you think I am guilty of the silly deterministic interpretation of the novel through the life of the author, you are wrong. My argument is, instead, based on the form, or poetics of the novel, as well as on an inserted novella, "The Captive's Tale", which tells about the slavery in Algiers — which is where he was held — and even mentions the author's second surname he adopted *after* his return to Spain, "Saavedra". The fictionality of that novella is clear from the fairy-tale outcome imagined, but the condition of slavery remains as its primary situation, from which the make-believe ending had liberated him.[2]

It is the poetics of the novel that is so "mad" that we can only relate it (rather than explain it in a causalistic sense) to an experience that, according to current psychiatry, must have been traumatising. This condition confronts us with the question of fiction and truth. As readers and, hence, public of the novel that this man wrote "post-" slavery, we are responsible for, first, understanding its relevance to our world, hence, to today; and, second, for doing something about it, which we can only do with the very limited means we have, *qua* public. Instead of dis-believing, we must acknowledge that there is no "post-" traumatic state as we casually call it in the medical phrase PTSD (Post-Traumatic Stress Disorder), because trauma does not live in time. The young soldier had no idea if and when he would ever get out during the five and a half years of his captivity. Imagine the feeling — or rather, the incapacity to have any. His parents did not have the money to redeem him. A few texts have been written by eyewitnesses or fellow slaves, describing the everyday life in the *baño,* the confrontations with cruelty and benevolence. But not much transpires about how the detained experienced their situation. We can only imagine. This is the first duty we have as public: to imagine what it is like.

Indeed, we need the imagination in the face of such un-representable events and situations. Captivity is not only a horrific experience, but the worst of it is, I would think, not knowing if there will ever be an end to it. Time loses its meaning. And it stretches endlessly. Into today. This harrowing temporality is at stake in the exhibition we have made, in confrontation with the temporal liberty offered to the visitors. This is, then, the second condition to be able to do something: temporal liberty so emphatic that it makes us imagine its opposite. To achieve this, I have asked for seating, benches, and/or chairs, in front of the video screens, as close as possible to these. Temporal freedom encourages a different relationship to visual art, something I began to experiment with in the 2017 exhibition I curated at the Munch Museum in Oslo, Norway. In the case of the stagnation of time in situations of trauma, there is an additional, more profound reason for this.[3]

In what follows I will, first, mention a few reasons why *Don Quijote* is such an emblematic case of the truthfulness of the lying about authorship and its responsibility. Needless to say, without going into the culture of lying tweets, I firmly reject the implied binary in the intolerable word "post-truth", and its abuse by systematic liars, and instead, I will argue that Cervantes' multiplication of authorship usefully insisted on the importance of a multiple imagination in his publics. Then, I will briefly present the exhibition of video work that engages and responds to the novel. A few examples from the video pieces will have to suffice to make my point concerning what kind of public is needed in the current state of the world, and why that has no use or even relevance for the idea that the question of truth-or-not is at issue in such imaginative cultural artefacts we call fiction, but that this does not make art less important, on the contrary. A different, important concern is at stake, for which artists must produce, construct, or "fabricate", and act upon a specific public, which is to respond by endorsing co-authorship.

Who wrote *Don Quijote?* Creative reversals

Probably the best-known, most frequently quoted sentence from Cervantes's novel is its first and worst lie:

> *... he buried himself in his books that he spent the nights reading from twilight till daybreak and the days from dawn till dark; and so from little sleep and much reading, his brain dried up and he lost his wits.*

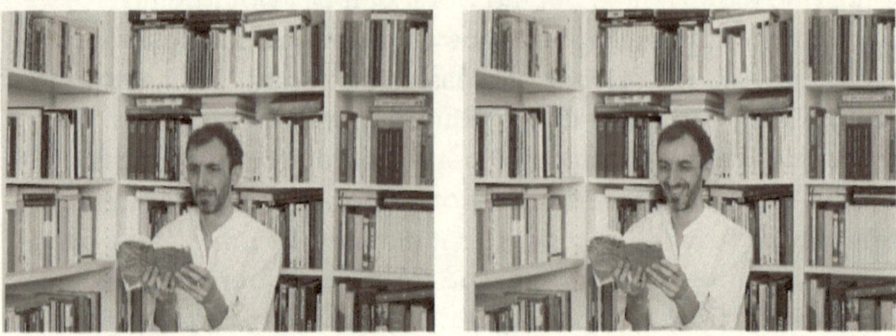

Figures 1 & 2. Don Quijote reading (Photos: Lena Verhoeff).

The implausibility of this 24-hours reading is one thing; but, the medically dubious result is what interested me most. The passage continues with a statement on fiction and its effect on readers, when it says:

> *He filled his mind with all that he read in them, with enchantments, quarrels, battles, challenges, wounds, wooings, loves, torments and other impossible non-sense; and so deeply did he steep his imagination in the belief that all that fanciful stuff he read was true, that to his mind no history in the world was more authentic.*[4]

With the exception of "enchantments" — a concept needed frequently to make the leaps in the (il)logic of the stories — these events called "nonsense" describe quite accurately the state of the world.

Since this "diagnosis" is itself so nonsensical, I take the liberty, contaminated by the idea of enchantment, to reverse it. Don Quijote did not go mad because of reading, but he fled into frantic reading because those other "authentic" events that constitute the world had made him mad. Reading was his self-curing strategy — a kind of auto-analysis. This reversal is a simple device to understand Cervantes's novel on a different mode than the laughable account of madness we can so easily dismiss while being amused by it. Instead, it is worth taking the future Knight Errant a bit more seriously, so that he is not a priori a stranger to us. In Chapter 9 of Part I, the narrator — called so to avoid projecting interpretations directly onto the author — has a lot to say about the authorship; statements that again seem highly implausible as well as relevant in its ostentatious confusions. This narrator who says "I" is, in fact, a listener — a member of the public to the story told by "the author"; already a fictional narrator, who could find no more records of the story. This merging of author and public seems highly relevant to me.

The tertiary narrator is the figure who, in the Alcaná of Toledo, found a manuscript he recognised as written in Arabic. The translation begins with "History of Don Quijote de la Mancha, written by Cide Hamete Benengeli, Arabic historian".

After the revealing statement that:

> *[I]f any objection can be made against the truth of this history, it can only be that its narrator was an Arab—men of that nation being ready liars, though as they are so much our enemies he might be thought rather to have fallen short of the truth than to have exaggerated.*

follows, however, the famous definition of the truth as "whose mother is history, rival of time, storehouse of great deeds, witness of the past, example and lesson to the present, warning for the future". Whereas the bad opinion of Arabs might be taken with a grain of salt, especially since its partiality blatantly contradicts the requirement of impassionate neutrality that follows right on its heels, the definition of truth is worth considering. *History as a rival of time* is a way of saying that the past and the present are not in chronological sequence, as I have argued at length in an earlier study. Witness, example, and warning are clear-enough aspects of the mixture of times in history to see the relevance of it.[5]

The Arab historian is well-enough appreciated later on, when the narrator says in Chapter 26:

> *... Cide Hamete Benengeli was a very exact historian and very precise in all his details, as can be seen by his not passing over these various points, trivial and petty though they may be. He should be an example to those grave historians who give us so short and skimped an account of events that we scarcely taste them, and so the most substantial part of their work, out of careless-ness, malice, or ignorance, remains in their ink-horns.*

So, the web of fictions and declarations of reliability is no longer logically understandable, but that doesn't matter. The point, as Jorge Luis Borges made clear in his "Pierre Menard, Author of the Quijote", is to take seriously what is written.[6]

The multiplication of narrators and the confusion between narrators and listeners or readers justifies the reversal I have proposed above and displaces the issue of truth from factuality to affect. It also allows the reversal of authorship and publics, who swap roles during all phases of the exhibition. If the novel's hero was so deeply hurt by the world, defined by the events mentioned, the only possible attitude towards "him" is empathy. And this is the ground on which I build my theory of a public not focused on historical truth — which is not a denial of it! — but on an attitude "for the world".

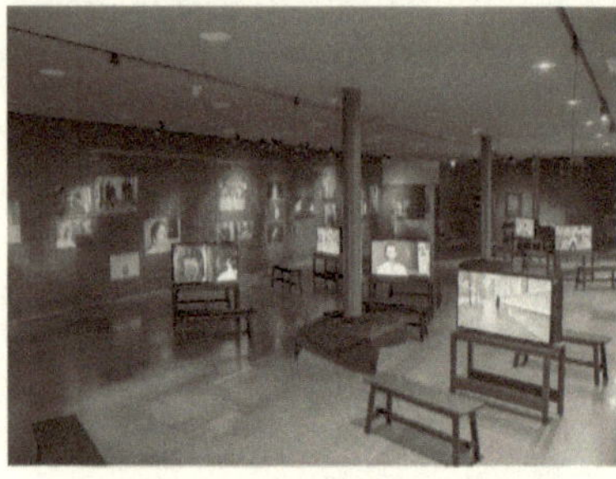

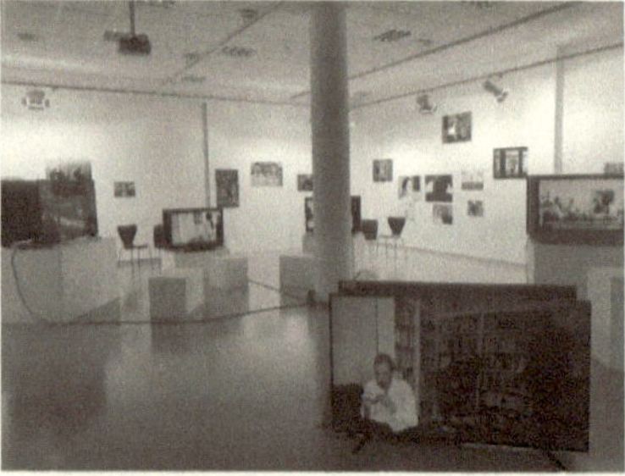

Figure 3. Installation in the Smålands Museum, Växjö, Sweden (Photo: Ebba Sund) & Figure 4. Installation in Facultad de Bellas Artes — Universidad de Murcia, Spain (Photo: Luz Bañón).

The point of this project is to bring the university, museums, and the publics together in a dialogue without mastery. The research question underlying it is this: how can art, museums, and theatre together help in the current situation of the world — mass migration, dictatorships, religious and nationalistic strife, destruction of the planet – to counteract violence's assault on human subjectivity, resulting in trauma? The question is examined through sixteen video pieces and thirty photographs, disorderly displayed and, by means of the benches, inviting visitors to sit, take time, and feel as if they are in a theatre rather than a museum. In their theatrical display, these videos and photos constitute an interdisciplinary case study, anchored in critical reflection and experimental art-making. The project deploys art in a revised museum practice in order to affect spectators: in this case, with the otherness of a socio-cultural state of violence-induced "madness". As mentioned, the tool is *empathy.* This term indicates "the capability to 'think in the mind of another', to anticipate the reactions of another human being".

This is not easy when that "other" is strange to us because of being "mad". If the public is willing, however, to bring empathy to madness, the figure of Don Quijote, the classical "mad knight" will be transformed into a "sad knight".[7]

As much as the author is pluralised, so is the public. On the one hand, museum visitors have an interest in art and in looking, a certain educational level and interests in common, as well as their common status as living today in the world as it is. On the other hand, no two people are exactly alike, and it is not the task of the museum to dictate what they ought to think. Instead, museums help to make people think, in a different way from the intellect only. This is why I have proposed to shift from *activist* art, useful as it may be in specific cases regarding specific issues, to what I call *activating* art: art that shakes up complacency, and makes people think on the basis of perception and affect, and perhaps changes their political opinions. But the pluralised author is not in a position to speak with a single voice. Hence, they do not dictate what readers think; they can "discuss" with them. This is what "fabricating publics" can mean in a positive, constructive sense. The key terms in this project — trauma, empathy, affect, time, and (preposterous) history — are the raw materials for that fabrication.[8]

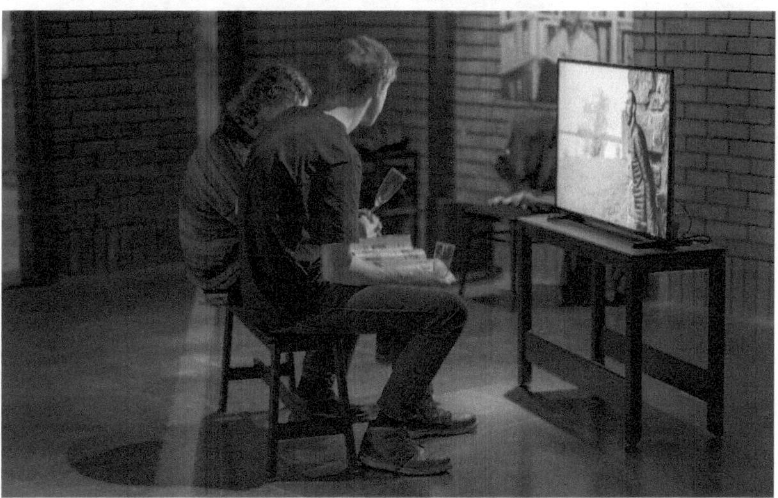

Figure 5. Two visitors watching the suspenseful moment when the Captive escapes (Photo: Ebba Sund).

The publics can be addressed by the collective authors. In the case of this video project, these comprise the actors, of which the pair of Mathieu Montanier (Don Quijote) and Viviana Moin (Sancho Panza) are particularly active as co-authors. Mathieu initiated the project, and he and I developed its first stages together. We sought to address the public on the basis of their autonomy-and-collectivity as one whole. We wanted to offer them a visible version of a fictitious "social life". The visuality of social life is a meaningful entrance into questions of what subjectivity is, how it can be perceived, and what this visibility tells us about human existence on the apparently shallow, yet so profoundly formative "stage" of interaction. Visual representations and interactions, sense-based presentations and absorptions shape the world as we see it. Images of desirable postures and faces, bodies and clothes, flickering colours of light, smiling and unsmiling faces fill our fantasies before we can even have any. Some of these images captivate us for a little bit longer than most; others fleetingly pass, but do not fail to leave their mark.[9] The publics inevitably rewrite the adventures and their stories.

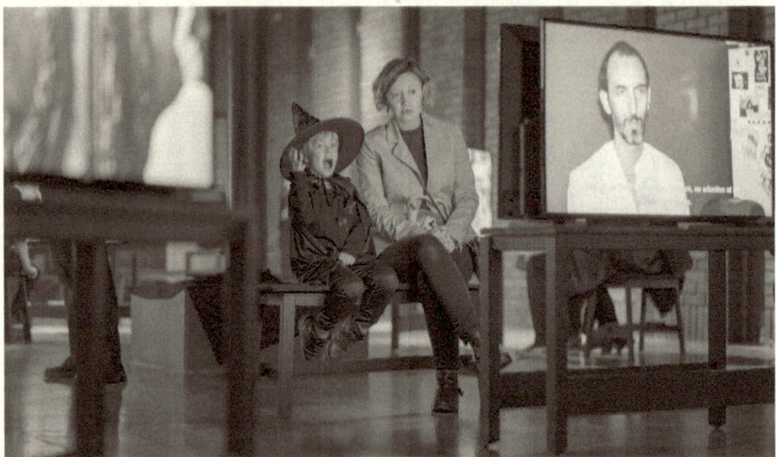

Figure 6. One laughing, the other looking concerned (Photo: Ebba Sund).

During the opening of the first exhibition in Sweden, I saw, for example, a five-year old boy adorned with a witch's hat (it was Halloween) rolling on the floor with laughter, whereas others, sitting on the same bench, alternated amusement with concern, when watching the episode "Pointless Altruism" (I refer to this scene in the following section). The different responses are significant in their pluralisation of the publics, who become co-authors; but, equally significantly, each one of the people sitting on that bench does respond: they are all affected, albeit in different ways, developing different moods. That freedom, but without indifference, is the point of art, and of this installation specifically. That is what I aim to demonstrate here.

Trying to help: How pointless is it?

The scene where Don Quijote roams around the city of Murcia in search of occasions to be helpful to people, matches the novel's primary tenor quite precisely, even if the setting and the situations are unapologetically contemporary. Going out and about with the desire to help as a motor, the Don Quijotes, old and new, keep misconstruing what they see. The hectic and pointless movements, more than the specific occasions, are what characterise the poetics of the novel and the exhibition that responds to it. The episode is described in the exhibition's catalogue as follows: "In spite of good intentions, 'doing good' is not so easy to define". This episode strings together a number of small actions of useless helping. In some sense, this scene can be seen as a summary or *mise-en-abyme* of the novel, including its falling apart in many different mini-episodes, which is its specific poetics that I consider to be compelled by trauma. Full of energy, always trying to assist others, it seems the hero can only be active and feel alive when acting on behalf of others, not himself. Don Quijote roams around in a world he does not understand. Out of the blue, he tries to help people who don't need his help. Everything he sees makes him worry. There is something hectic about his behaviour. His body language conveys madness, despair, and a feeling of un-belonging. Yet, the situations he aims to repair appear quite ordinary, even if they do have an edge of danger.

He addresses and even physically attacks a man who seems to be risking his baby by being distracted talking on the telephone (Figure 7). A woman walking a dog is told that the dog is not comfortable on the leash; Don Quijote tries to take the leash from her. When cleaning up garbage that overflows the bins, he accidentally makes the mess worse. When he is protecting from more damage someone who has been wounded, he hurts the injured young student. Entering a clothing store, he tries to rearrange the display, as if he knew better what is appealing to shoppers. And more. What is normal? What is mad? Go figure!

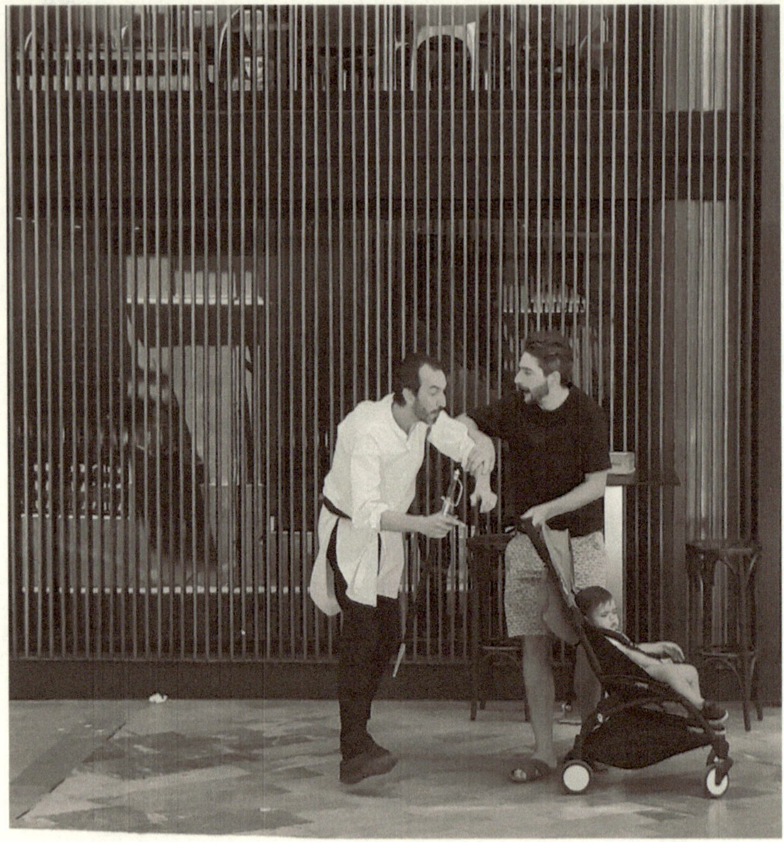

Figure 7. Don Quijote intervening, interfering (Photo: Luz Bañón).

Although the idea, from the novel, is that the Knight's attempts to help others are pointless, this doesn't mean that they really and always are. True, there is not much in the novel or in the videos that results in a meaningful improvement of situations. But the underlying claim, that the social fabric needs to shed its indifference, stays more or less valid. In the episode "Pointless Altruism" that I am describing here, for example, we see the knight in an urban setting, addressing, but in fact, aggressing passers-by who do something Don Quijote considers dangerous. His mistaken judgments bring those he tries to help to the edge of victimhood. The baby begins to cry, the dog to bark, and the wounded student flinches when Don Quijote rudely takes him to his side to protect him from further damage. The problem this episode foregrounds is the unbalanced relationship between individual and social existence. Don Quijote takes on such arbitrary issues because his sense of self and of his mission overrules the obvious fact that those others he tries to help have their own individualities. The scene also makes us reflect on the fine line between charitable and aggressive behaviour. Imposing his view of the situations, he becomes dictatorial.[10]

Part of the problem, both of the Knight Errant and of the readers of the novel and the viewers of the videos, is the absence of a clear "yes/no" answer to the dilemmas posed. And that was, precisely, one of my motivations to take this project on. How can such a work of "cultural heritage", admired and, yet, ridiculed both by the laughing readers and the "post-modern" filmmakers, do anything useful for the social world? Only in watching and listening to the episode "The Great Peacemaker *versus* Money" (also in an urban setting — Paris) does it become appealing to reflect on the things he says and the value of his intervention. Clearly, there is, again, no "good/bad" dilemma, and no answer is likely to satisfy. The scene critiques yet remains steeped in capitalism. It demonstrates that you cannot step out of ideology, even if challenging it from within remains an area where small changes can happen and can accumulate.

Ridicule, however, is not a helpful answer to what one considers "mad". This is why most films based on Cervantes's novel are hard to watch. Orson Wells didn't manage to finish it. All others stop somewhere arbitrary. The first problem is linearity; this is precisely what traumatisation destroys. The second problem is mockery. I am especially referring to Terry Gilliam's 2018 film, *The Man Who Killed Don Quixote.* After fifteen years he finally pulled it off, but at the cost of a "geronto-phobic", or "agist" caricature of the hero as an old man. This produces a remoteness from the public. Such distancing cannot produce the empathic publics our project attempts to solicit.

Another scene that focuses on pointless attempts to help is, in fact, based on an impressively radical feminist episode of the novel. We titled the scene "Woman as Anti-Suicide Bomb". The title is a quote from, and an homage to, Françoise Davoine, who uses that phrase in her book on *Don Quijote.*[11] In the incident presented in this scene, good will, bad faith, and other social attitudes are put to the test. This episode, based on Chapters 12–14 of Part I, questions the hysterical reaction to amorous rejection — the idea that some men won't take "no" for an answer. It is an early-modern scene of patriarchy versus feminism — again, of course, without using that term —, which, at the same time, recalls contemporary sexual pressure. The woman opposes a firm "'no' means 'no'" to the young men's fury about her rejection of their deceased friend.

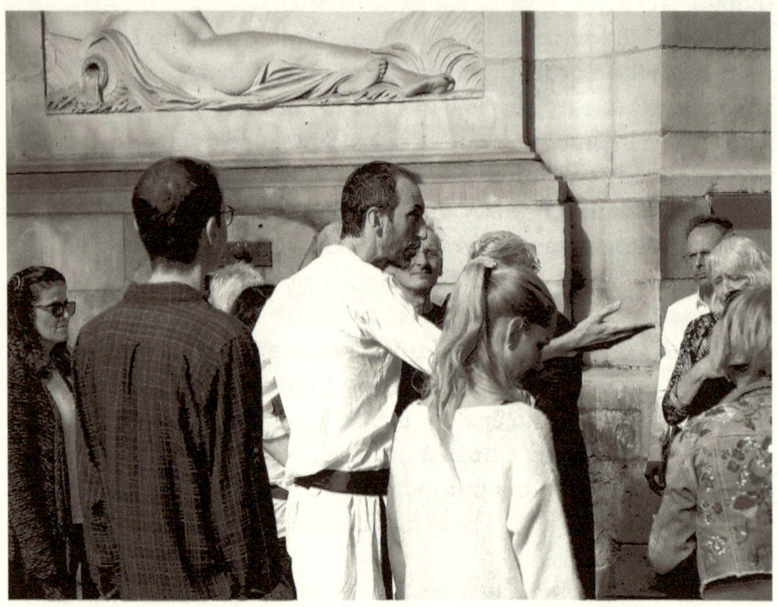

Figure 8. Don Quijote haranguing the crowd about social values (Photo: Toni Simó Mulet).

In addition to being the beginning of Don Quijote's pointless altruism, this scene explores the idea — which might seem contemporary but is already clearly and explicitly expressed in Cervantes's novel — that Marcela doesn't need Don Quijote's help, and that men who try to be "good guys" still reproduce some of the masculinist pitfalls in their interactions with women. The young men who lost their friend to suicide grieve, then project their sorrow on the woman when she accidentally comes upon the burial. "Because" she is beautiful, they are contaminated by their dead friend's despair, and blame it on her. To foreground this "inter-temporality", the audible dialogue consists entirely of quotations from the novel, except for the phrase "no es no".[12]

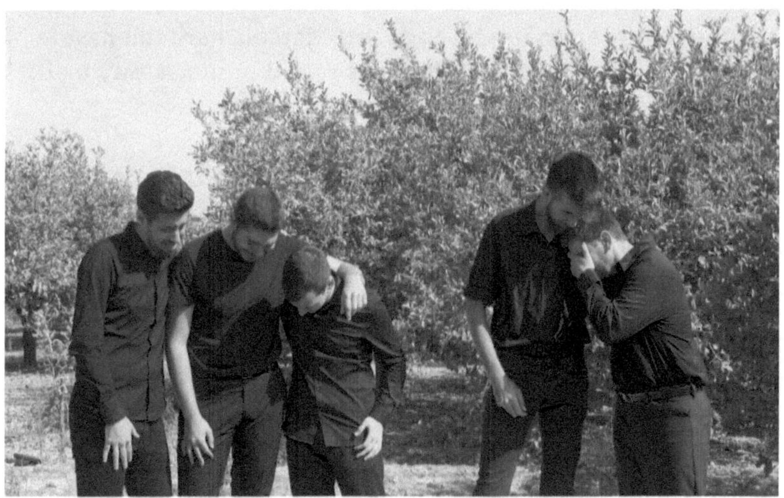

Figure 9. Mourners weeping for the suicide of Crisóstomo (Photo: Mar Sáez).

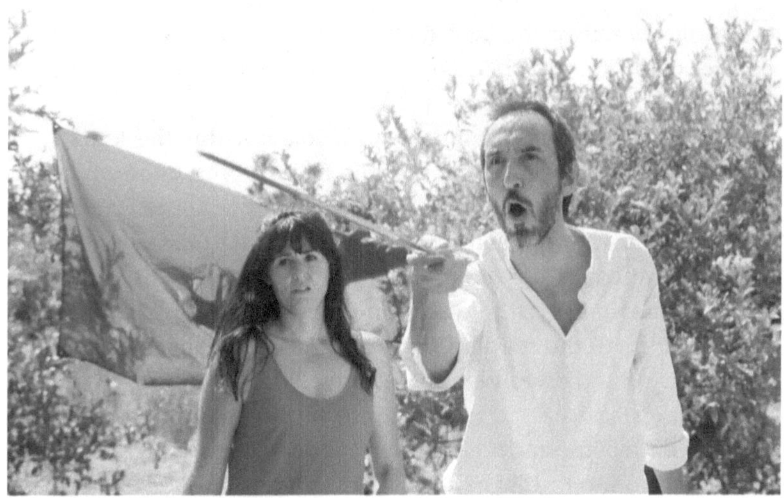

Figure 10. Don Quijote defending Marcela (Irene Villascusa) (Photo: Mar Sáez).

The moment Don Quijote hears the laments and the scolding of the friends of the diseased, he decides to intervene. The young woman Marcela, however, can defend herself. Her speech asserts her right to independence and is convincing enough. The suicide is indicted, not the woman who chooses to live her own life. But Don Quijote's desire to feel important in his crusade against injustice leads to his being contaminated by the wrong-headed mourners when he adds that Marcela may still have some guilt. He says, brandishing his sword in such a way that viewers might feel concern for Marcela's safety:

> *Let no man, of whatever state or condition, dare to follow the fair Marcela, under pain of incurring my most furious indignation! She has shown with clear and sufficient argument that she bears little or no blame for Crisóstomo's death, and how far she is from yielding to any of her [suitors'] desires. Wherefore it is right that, instead of being pursued and persecuted, she should be honoured and esteemed by all good men in the world, for she has proved that she is the only woman living with such pure intentions.*[13]

But rather than being grateful for his help, Marcela pushes him out of the way. This is the most feminist moment of the novel, impressively so, and deserved to be foregrounded. At the same time, for us as co-authors in the act of fabricating a committed public, we used the brandishing of the sword as the small gloss of caution to not be taken in by the "good guy" talk of Don Quijote.[14]

Is helping others pointless, then? It is not. But, the kind of help cannot be of the kind that the Knight, who knows how people should behave, offers, or rather, imposes. The confrontation with this condescending attitude is one of the elements of our attempt to make the publics think, including their own experiences. Instead, the question the novel opens and keeps reiterating in all the mad episodes, is that of a help that is not of superiority, but of involvement. An involvement that is emotional, reflective, and bound to what one can see, all at once. I still find "empathy" the most suitable concept to express this. It is in empathy that the public can be fabricated as one that both preserves its plurality, the singularity of each of its members, and the lived concern for the collectivity in which and with which we each live. Hence, without either the voyeuristic immodesty or the dilution of the concept, the most extreme, most problematic, and yet, most urgent situation a public can be enticed to address with empathy is trauma. This is an enticement, no more; not an "empathy-machine".[15]

Trauma needs empathy

My initial interest in "doing something" with Cervantes' novel, in addition to my eagerness to work with the brilliant actor who proposed the project, was the reversal I mention at the beginning of this essay. My firm conviction is that Cervantes, traumatised by his slavery, and, as a literary author and a believer in the healing power of reading, experimented with the reversal. The resulting novel is "mad": it carries not only the traces of the absurdity and madness that suggest the inevitably traumatic state in which its creator must have been locked upon his return to Spain, as transpired in the stories told, but also — even primarily — in the novel's poetics; it also foregrounds this consequence of war and captivity in the madness of its literary form. The sheer-endless stream of "adventures" makes all film adaptations more or less hopeless endeavours. One can barely read, let alone watch all those pointless attempts to help others, the repercussions of which involve cruelty and pain. Repetitiveness overrules narrative.

Not in order to "match" this as a "faithful adaptation", but to experiment with it in the belief that a brilliant author like Cervantes knew what he was doing with this mad poetics, we endeavoured to make an audio-visual work based on this novel. Thus, through being touched by the installation's form, viewers can learn from it in order to deal with their own experiences of the violence contemporary society can generate — their own as well as those hinted at by others in their surroundings. This attempt to make art exhibitions work for the social problem of reconnecting the excluded "mad" aims to repair what Cervantes called in the Prologue to his last novel, *The Travails of Persiles and Sigismunda* (1617), repairing the "broken thread" of memory. There is no more adequate and poetic articulation of what it means to be traumatised. This metaphor is quoted in the scene "Narrative Stuttering".[16]

In this project, therefore, the attempt is to present, but not re-present trauma. For this purpose, it is imperative to distinguish between three aspects of trauma: its cause; the situation or state that the cause produces; and the possibility of helping people suffering from it to come out of it. This distinction can be formulated succinctly as follows: *violence* is an event that happens; *trauma* is a state that results from the violence; *empathy* is an attitude that enables. The subjects of these three aspects are different: the violence has a doer (culprit, perpetrator);

the traumatised subject is the victim of, and is captured by it; and the subject of empathy is the social interlocutor, who can potentially help to overcome it. In the case of this project, it is the visitor who is the primary target of the exhibition; its interlocutor, and the interlocutor of the fictional figures brought to life. This exhibition aims to activate visitors to become such empathic subjects. The display is meant to have performativity. This is what the project aims to solicit in the publics it attempts to produce.[17]

Confusions and ethical problems threaten attempts to show such horrid acts of violence. In our exhibition we do not show these acts. A solicitation of feel-good identification potentially leading to "trauma envy" always lurks and is utterly unhelpful, even ethically problematic. So does the risk of voyeurism, as we know from Adorno's caution against it. Davoine writes in her book on *Don Quichottee:* "Cervantes doesn't try to arouse visions of horror for voyeuristic readers".[18] One moment where violence occurs in our videos is when a traumatised young man acts his madness out, but that is already as a consequence of earlier violence. This is in the episode "The Failure of Listening". The young Cardenio's attacks on his interlocutors are responses to the latter's failure to allow him to speak without being interrupted, whereas this is a condition he states prior to begin his story. This is where trauma can be encountered by empathy. This happens in a counterpart to this episode — the one where Don Quijote is listening to witnesses who are deeply involved in contemporary situations of refugees. There, he is able to be sensitive and forget his own obsessions. This scene, "Testimonial Discourses", acutely updates the traumatic events in the other scenes, so that visitors are alerted by the recognisable stories to the actuality of the issues Cervantes was able to draw out from his own life experience, with the help of his imagination.

One of the episodes where the traumatised is visible is what I have titled "She, Too", in an obvious allusion to the contemporary situation. A young woman, the beloved daughter of a slave owner — the one who "owns" the captive —, is so jealously guarded by her father that she is never allowed to go out. As a consequence, she, too is a captive; and thus, she, too is traumatised. This was my interpretation of the role of Zoraida in the novel. She is shown roaming around the castle, looking at pictures that show freedom, and she ends up contacting the slaves working outside, whom she almost envies. In the novel, she helps the Captive and his friends to escape, thus escaping herself, too. In this scene, I tried to give that offer of help a reason beyond the romantic interest in someone she cannot know other than seeing him from afar.

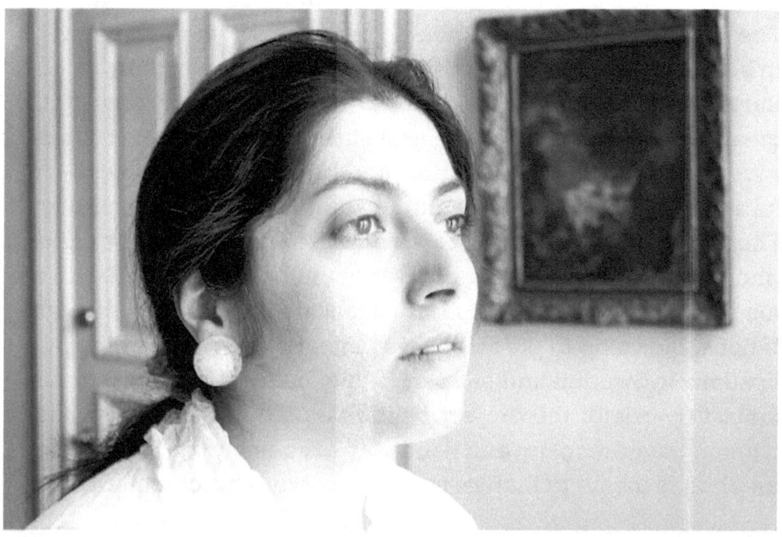

Figure 11. Zoraida (Nafiseh Mousavi) looks out with frustrated longing, or in catatonic stupor (Photo: Ebba Sund).

The publics the exhibition is aimed at fabricating can respond to this, changing their attitudes from the usual consequence of visual abundancy, the worst social ill, which I have been countering all through my working life: indifference. The many representations of traumatogenic events in the electronic media generate a forgetting of their historical and psychological impact. The far-too-many, the surplus, is produced by, and produces *consumption.* Through their graphic explicitness and their recurrent appearance, these pictures are confined to historical insignificance, even oblivion. As mentioned above, our project designs an intervention in that cultural attitude, by inflecting *activist* art into *activating* art, public-oriented, for a more general change of attitude. The case is made for a community-creating effect of art that helps to repair the broken social bond that has resulted in trauma. The traumatised person is alone, and not even able to (fully) remember the horror that caused the state of trauma. As a result, they are even alone within themselves. If anything can be done to help such victims exit their paralysing state of stagnation, it must be done through reducing that double loneliness. This is a social task for which everyone is qualified.[19]

This brings me to a final point about authorship and publics together. This compels me to give the last word to a visual artwork, made by a young artist to whom I wish to pay homage. The photo we have used as the poster image of the exhibition in Växjö, Sweden, and the cover image of the publication, is the most brilliant visual, bi-dimensional presentation of what it means to be traumatised. The artist has captured the essence of what is at stake in the exhibition, and beyond that, in the cultural field where we need multiple authors and multiple publics. Therefore, I want to introduce artist Ebba Sund, an independent photographer based in Sweden, as co-author of the work. She has made many beautiful photographs that we have used in the exhibitions, blown up to the size of wall images, as well as in the publications in Sweden and in Spain. While many others have done fantastic work for the project, which would not have come about without the generosity and solidarity of so many, my focus here on Ebba is meant to make a statement about this co-authorship.

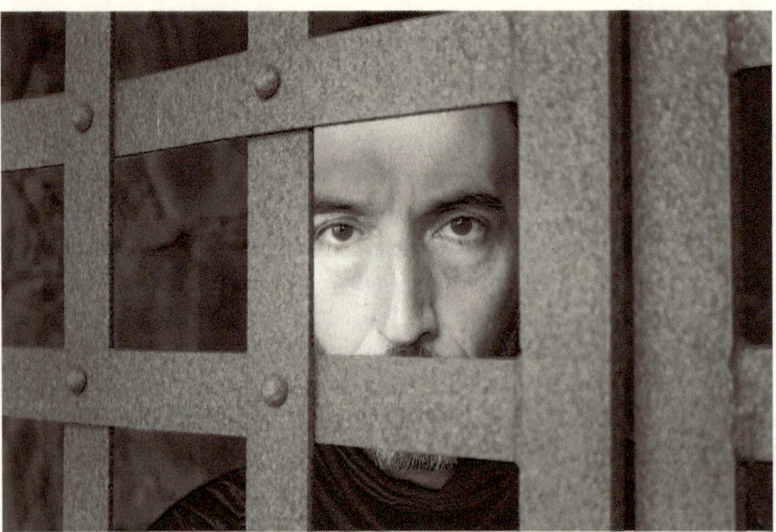

Figure 12. The Captive cannot speak with his mouth, only with his eyes (Photo: Ebba Sund).

The imploring eyes speak for themselves. The actor, the photographer, the site with its grid: they all co-authored this project. It is my dearest wish that the exhibition of this material in various places will slightly shift the media-generated social indifference towards an attitude of empathy; not only with the actual millions of slaves living in the contemporary world, but with all forms of unfreedom. And if the image is, in addition, so utterly beautiful, this only goes to show that art and its aesthetic power matter. Which is why it can and must fabricate publics that are willing and able to connect to others. That is the truth: one that bears no "post-", and mobilises the alleged "lies" of fiction to exert art's agency.

Notes

1. Mieke Bal, "Challenging and Saving the Author, For Creativity", Vesper. *Rivista di architettura, arti e teoria | Journal of Architecture, Arts & Theory: Materia-autore | Author-Matter,* no. 2 (Spring–Summer 2020): 132–149. I won't go further into my argumentation here.

2. On Cervantes's captivity, see María Antonia Garcés, *Cervantes in Algiers: A Captive's Tale* (Nashville, Tennessee: Vanderbilt University Press, 2002). For a contemporary witness, María Antonia Garcés (ed.) *An Early Modern Dialogue with Islam: Antonio de Sosa's Topography of Algiers* (1612). trans. Diana de Armas Wilson (Notre Dame, IN: University of Notre Dame Press, 2011).

3. There, my decision to put benches in front of the paintings by Munch was to put his art on a par with the included videos of *Madame B,* since video is a time-based art. See the book that accompanied the exhibition (2017), in which I explain the importance of time in exhibitions: Mieke Bal, *Emma & Edvard Looking Sideways: Loneliness and the Cinematic* (Oslo: Munch Museum/Brussels: Mercatorfonds: Yale University Press, 2017). *Madame B* is a 19-channel video installation from 2014 that I made with Michelle Williams Gamaker.

4. I use the most accessible translation by J. M. Cohen and will refer to the short chapters rather than page numbers, to enable readers who use other editions to follow the argument. I have checked all translations against the Spanish and French, and only modify them when they affect my argument. This quote is on the second page of the first chapter. See Miguel de Cervantes Saavedra, *The Adventures of Don Quixote,* trans. J. M. Cohen (London: Penguin Books Ltd., 1950).

5. This argument led to the concept of "preposterous history", which I defended with the help of contemporary artists "revising" Caravaggio (1999). See Mieke Bal, *Quoting Caravaggio: Contemporary Art, Preposterous History* (Chicago, IL: University of Chicago Press, 1999).

6. Borges's short story "Pierre Menard, Author of the Quixote" is in *Labyrinths* (Harmondsworth: Penguin, 1962), 62–71. See my *Narratology* for a brief analysis: Mieke Bal, *Narratology: Introduction to the Theory of Narrative* (Toronto: University of Toronto Press, 2017 [1982]).

7. A useful collection on the ups and downs of empathy is Aleida Assmann and Ines Detmers (eds.) *Empathy and its Limits* (London: Palgrave, 2016). The quote is taken from the first page of the introduction. Jill Bennett connects empathy and trauma when she discusses Dominick LaCapra's and other reflections in a subtle and accessible manner in her book *Empathic Vision: Affect, Trauma, and Contemporary Art*

(Stanford, CA: Stanford University Press, 2005).

8. Affect and trauma are contentious and much-abused terms. For the most lucid analysis of the reasons for the unrepresentability of trauma, see Ernst van Alphen, "Symptoms of Discursivity: Experience, Memory, Trauma", in *Acts of Memory: Cultural Recall in the Present,* eds. Mieke Bal, Jonathan Crewe, and Leo Spitzer (Hanover NH: University of New England Press, 1999): 24–38. This author also wrote a most helpful overview of the issues of affect: Ernst van Alphen, "Affective Operations of Art and Literature", in *RES: Journal of Anthropology and Aesthetics* 53/54 (Spring/Autumn 2008): 20–30. This can be usefully complemented with Eugenie Brinkema's book *The Forms of the Affects* (Durham, NC: Duke University Press, 2014).

9. For a classic on the visibility of social life, see Erving Goffman, *The Presentation of Self in Everyday Life* (New York: Doubleday, 1956). For a psychanalytic account, Christopher Bollas, *The Shadow of the Object: Psychoanalysis of the Unthought Known* (New York: Columbia University Press, 1987). On theatricality as "a critical vision machine", see Maaike Bleeker, *Visuality in the Theatre: The Locus of Looking* (Basingstoke: Palgrave Macmillan, 2008); and Maaike Bleeker, "Being Angela Merkel" in *The Rhetoric of Sincerity,* ed. Ernst van Alphen, Mieke Bal, and Carel Smith (Stanford, CA: Stanford University Press, 2009): 247–262. In these publications, Bleeker demonstrates, through detailed analyses, how productive the concept of theatricality can be for a political art that is not bound to a political thematic.

10. Here, I quote fragments from the brochure and accompanying book, both published by Niklas Salmose and his company Trolltrumma, in Växjö, Sweden, where the book can be ordered at niklas@trolltrumma.se. I am deeply grateful to Niklas, a colleague at Linnaeus University, who has done more for the project than I can possibly acknowledge.

11. Françoise Davoine, *Don Quichotte, pour combattre la mélancolie* (Paris: Stock, 2008).

12. For a very relevant volume on the shamefulness of such masculinist behaviours, see Ernst van Alpen (ed.) *Shame! and Masculinity* (Amsterdam: Valiz, 2020).

13. Cervantes, *The Adventures of Don Quixote,* Part 1, Chapter 14 (emphasis added).

14. The sword, quite an impressive one, served excellent purposes in the project. It was given to me by the University of Helsinki at the occasion of an honorary doctorate in 2019. I am grateful to Professor Kirsi Saarikangas for initiating this great event.

15. See Davoine, *Don Quichotte;* also Françoise Davoine and Jean-Max Gaudillière, *A bon entendeur, salut!: face à la perversion, le retour de Don Quichotte* (Paris: Stock, 2013). Davoine has had a deep influence on my video work from the first "theoretical fiction", on the 2011 film and video project, *A Long History of Madness,* where she collaborated in many ways, including playing her own character. This film is based on her book *Mother Folly: A Tale,* trans. Judith G. Miller (Stanford, CA: Stanford University Press, 2014 [or 1998]). I also thank London-based Peruvian scholar Luis Rebaza-Soraluz for insisting on the feminist aspect of the novel. The creation of the episode "She, Too" was also influenced by that insistence.

16. On the problematic notion of "faithful adaptation", see the classical essay by Thomas Leitch, "Twelve Fallacies in Contemporary Adaptation Theory", *Criticism 45,* 2 (2003): 149–171.

17. A further study on performativity would best begin with the original: J. L. Austin, *How to Do Things with Words* (Cambridge, MA: Harvard University Press, 1975 [1962]). Of the many discussions of performativity, I consider Jonathan Culler's to be the most lucid: "The Performative", in *The Literary in Theory* (Stanford, CA: Stanford University Press, 2007): 137–165.
A brilliantly ground-shifting recent text focusing on trauma is Ernst van Alphen, "The performativity of provocation: the case of Artur Zmijewski", *The Journal of Visual Culture* 18:1 (2019): 81–96.

18. Davoine, *Don Quichotte,* 93.

19. Davoine's life work has always been devoted to precisely that: repairing the broken social bond. Our project with her mentioned above, *A Long History of Madness* (2011), which was based on her theoretical fiction *Mother Folly* (2014), also avoided to represent trauma, while centrally presenting it. In my struggle against indifference, I have written about "an ethics of non-indifference" in Mieke Bal, *Loving Yusuf: Conceptual Travels from Present to Past* (Chicago, IL: University of Chicago Press, 2008).

References

Alphen, Ernst van. "Symptoms of Discursivity: Experience, Memory, Trauma". In *Acts of Memory: Cultural Recall in the Present,* edited by Mieke Bal, Jonathan Crewe, and Leo Spitzer, 24–38. Hanover, NH: University of New England Press, 1999.

Alphen, Ernst van. "Affective Operations of Art and Literature". *RES: Journal of Anthropology and Aesthetics* 53/54 (Spring/Autumn 2008): 20–30.

Alphen, Ernst van. "The performativity of provocation: the case of Artur Zmijewski". *The Journal of Visual Culture* 18:1 (2019): 81-96.

Alpen, Ernst van (ed.) *Shame! and Masculinity.* Amsterdam: Valiz, 2020.

Assmann, Aleida and Ines Detmers (eds.) *Empathy and its Limits.* London: Palgrave, 2016.

Austin, J. L. *How to Do Things with Words.* Cambridge, MA: Harvard University Press, 1975 [1962].

Bal, Mieke. "Challenging and Saving the Author, for Creativity". *Vesper. Rivista di architettura, arti e teoria | Journal of Architecture, Arts & Theory: Materia-autore | Author-Matter,* no.2 (Spring-Summer 2020): 132–149.

Bal, Mieke. *Don Quijote: Sad Countenances,* edited by Niklas Salmose. Växjö, Sweden: Trolltrumma, 2019.

Bal, Mieke. *Emma & Edvard Looking Sideways: Loneliness and the Cinematic.* Oslo: Munch Museum/ Brussels: Mercatorfonds: Yale University Press, 2017.

Bal, Mieke. *Narratology: Introduction to the Theory of Narrative.* Toronto: University of Toronto Press, 2017 [1982].

Bal, Mieke. *Loving Yusuf: Conceptual Travels from Present to Past.* Chicago, IL: University of Chicago Press, 2008.

Bal, Mieke. *Quoting Caravaggio: Contemporary Art, Preposterous History.* Chicago, IL: University of Chicago Press, 1999.

Barthes, Roland. "The Death of the Author". *In The Rustle of Language,* translated by Richard Howard, 49–55. New York: Hill and Wang, 1986.

Bennett, Jill. *Empathic Vision: Affect, Trauma, and Contemporary Art.* Stanford, CA: Stanford University Press, 2005.

Bleeker, Maaike. *Visuality in the Theatre: The Locus of Looking.* Basingstoke: Palgrave Macmillan, 2008.

Bleeker, Maaike. "Being Angela Merkel". *In The Rhetoric of Sincerity,* edited by Ernst van Alphen, Mieke

Bal, and Carel Smith, 247–262. Stanford, CA: Stanford University Press, 2009.

Bollas, Christopher. *The Shadow of the Object: Psychoanalysis of the Unthought Known.* New York: Columbia University Press, 1987.

Brinkema, Eugenie. *The Forms of the Affects.* Durham, NC: Duke University Press, 2014.

Cervantes, Miguel de Saavedra. Versions used:

Spanish: Cervantes, Miguel de. Don Quijote de la Mancha, puesto en castellano actual integral y fielmente por Andrés Trapiello. Barcelona: Ediciones Destino, 2016 [1605, 1615].

French: Cervantès, L'ingénieux Hidalgo Don Quichotte I et II. Edition nouvelle présenté, traduit et annoté par Jean Canavaggio. Paris: Gallimard, 2001.

English: Cervantes, Miguel de Saavedra. *The Adventures of Don Quixote,* translated by J. M. Cohen. London: Penguin Books Ltd., 1950.

Culler, Jonathan. "The Performative". In *The Literary in Theory,* 137–165. Stanford, CA: Stanford University Press, 2007.

Davoine, Françoise. *Mother Folly: A Tale,* translated by Judith G. Miller. Stanford, CA: Stanford University Press, 2014 [1998].

Davoine, Françoise. *Don Quichotte, pour combattre la mélancolie.* Paris: Stock, 2008.

Davoine, Françoise and Jean-Max Gaudillière. *A bon entendeur, salut!: face à la perversion, le retour de Don Quichotte.* Paris: Stock, 2013.

Foucault, Michel. "What is an Author?" In *Textual Strategies: Perspectives in Post-Structuralist Criticism,* edited by Josué V. Harari, translated by Donald Bouchard and Sherry Simon, 141–160. Ithaca, GR: Cornell University Press, 1979.

Garcés, María Antonia. *Cervantes in Algiers: A Captive's Tale.* Nashville, Tennessee: Vanderbilt University Press, 2002.

Garcés, María Antonia (ed.) *An Early Modern Dialogue with Islam: Antonio de Sosa's Topography of Algiers* (1612), translated by Diana de Armas Wilson. Notre Dame, IN: University of Notre Dame Press, 2011.

Goffman, Erving. *The Presentation of Self in Everyday Life.* New York: Doubleday, 1956.

Leitch, Thomas. "Twelve Fallacies in Contemporary Adaptation Theory". In *Criticism 45*: 2 (2003): 149–171.

Birth of the Cool

Charlie Gere

2016 did not start well. In January, I was backing the car out of the drive when I heard that David Bowie had died. I had to stop and let the news sink in. For people of my generation, Bowie was a touchstone. Like W. H. Auden's memorial description of Sigmund Freud, he was "a climate of opinion" as much as a person. He dominated my youth and that of my generation in a manner unlike any other public figure. However, much as I love Bowie's music — above all, the albums he released around the mid-1970s — I became suspicious of his elevation to a kind of secular sainthood. Bowie was no saint, especially in the period within which he was making his best music. For example, his sexual exploitation of very young girls during that period does not look good in light of contemporary sex scandals and the #MeToo movement. Moreover, some of his political posturing in the 1970s was both reprehensible, and strangely prophetic of our current situation. In retrospect, his death could be seen as an omen of the momentous changes the year of his demise would bring, not least because he himself seemed to anticipate them.

Twice in 2016 people woke up to find themselves in a different world to the one in which they thought they lived. In the United Kingdom, this happened on the morning after the EU referendum on the UK's membership of the European Union on 23 June. In the United States, it was after the general election on 8 November. In both cases, the result was — for many people, at least — highly surprising and disturbing. Not only were these results unexpected, but they also revealed an entirely transformed political and cultural landscape to that of progressive liberal democracy. Something momentous had occurred — a profound cultural shift. It was as if, to paraphrase Virginia Woolf, on or about June or November 2016 human character changed. Woolf had made her original remark, dating such a change to December 1910, to reflect the massive transformations brought about by an accelerating modernity in the early-twentieth century, transforming the world through scientific and technological change.[1]

Clearly, Woolf's specificity about the date was a little flippant, and the changes she alluded to had been happening for quite a long time before then (although, she was not quite as specific as Charles Jencks, who dated the beginning of post-modernism to 3:32pm on the 15 July 1972 — the precise moment of demolition of the exemplary modernist Pruitt-Igoe housing project in St Louis, Missouri, that heralded the death of modernist architecture).[2] What Woolf meant, perhaps, was that it was only then that these changes became obvious. There are clear parallels with our current situation, though the transformations, in our case, are those brought about by digital technologies and, in particular, social media networks.

What was revealed in the morning light of those awakenings of 2016 was a radically new digital culture of algorithmically-driven creative destruction, of Big Data used in the service of social engineering; of election hacking through social media by hostile powers, of virulent and vicious hatred; bullying and violence spread via the same networks; and of the return of repressed forces of Fascism, racism and other similar elements. The social media networks once regarded as benign ways of making connections and enabling self-expression, turned out to be far more sinister. Facebook's failure to address its role in the spreading of fake news and other abuses exemplified this. What all this seemed to suggest was almost an alternative culture, based around chatrooms, blogs, Facebook timelines, and so on. This new culture both operated outside the normal media such as television, film, and newspapers, and also parasitised on them, to the extent that it was hollowing out such media, much as parasites can do to their host organisms.

Unsurprisingly, perhaps, such media and other official conduits of knowledge and discourse such as universities, show little evidence of really understanding what is happening to them. Or, perhaps, the truth is they understand perfectly well, but this understanding does not change their practices and modes of being. In effect, the traditional media are dying, or becoming so marginal as to be irrelevant, or so transformed as to be unrecognisable. An early example of this is the failure of the music industry to engage with the challenges of streaming. These phenomena are what constitute our culture now. We are currently experiencing a massive shift in our discourse networks, with the increasing ubiquity of networks and the World Wide Web. The Web is more, far more, than merely a communications tool (as if anything is *merely* a communications tool). It changes everything. All art, literature, theatre, film, now takes place in a context entirely determined by our networked, mediated culture, by our social media. The houses of culture are now simply façades, hollowed out to accommodate the data farms of contemporary digital culture.

Back in 1999, Bowie predicted the cataclysmic effect of the Web in an interview with a sceptical Jeremy Paxman. Bowie was an early adopter, predictably perhaps, and by this time had already founded BowieNet, his own internet service provider (ISP). The interview is an indicative encounter between a fairly complacent Oxbridge/mainstream media attitude from Paxman, and a brilliantly intuitive understanding from Bowie about what is happening. It exemplifies why the mainstream media establishment has not yet understood what is going on. In Bowie's words, "I think the potential of what the internet is going to do society — both good and bad — is unimaginable. I think we're actually on the cusp of something exhilarating and terrifying". Paxman replied saying, "It's just a tool, though, isn't it?" "No it's not", stated Bowie. "It's an alien life form. Is there Life on Mars? Yes, it's just landed here".[3]

Bowie was right, of course, as no one can seriously doubt after Trump, Brexit, Gamergate, "involuntary celibate" ("incel") massacres, ISIS, the scandals of Facebook, Cambridge Analytica, Russian election hacking, "post-truth", etc. All these are the direct results of the massive techno-social developments described above. At a less immediate level, there is the whole world of selfie culture, cyberbullying and so on, as well as the ubiquity of the meme as a cultural response, the importance of chatrooms, the use of algorithms, artificial intelligence, and so on.

The kind of ephemeral cultural production found on the Web such as LOLcats, Doge, Vaporwave and all the other memes are a form of avant-garde folk practice. Many of these phenomena are very funny and incredibly vital. But many are increasingly employed in the service of reactionary ideas, epitomised by the Pepe the Frog cartoon character appropriated by the far right as a meme, and the extent to which the devil has all the best tunes. In the digital realm, the discourse on the right is often horrendous, hateful, virulently misogynist, racist, anti-Semitic, exceptionally violent, and nihilistic; but it can also be very funny. Recent media exposure of incel culture revealed a host of comic figures and ideas such as the Chads and Stacys. Chad Thundercock is the name given to sexually successful men by so-called incels, men who believe they are being deprived of the sex they deserve by women whom they characterise in the most appallingly denigratory terms. He is the preternaturally handsome man who casually fucks the girl you have had a crush on for months, and then just as casually abandons her. Stacy is her female equivalent. I find the names very funny, which makes me extremely uncomfortable, knowing that incels have committed mass murder in the name of their pathological self-loathing. This is an extremely important issue — the capacity of alt-right Web discourse to play effectively with humour and irony in a highly nihilistic manner, and an abandonment of any higher meaning or purpose. The role of irony and humour in alt-right discourse is of considerable importance and has massive political and social implications.

All this is a profoundly political issue; one of the most important, and chilling principles underlying the exploitation of social media by the alt-right is the idea that, in Andrew Breitbart's words, "politics is downstream from culture".[4] It was this realisation that led Breitbart's colleague and successor at Breitbart News, Steve Bannon, to see the value of fashion forecasting for political ends. In the words of Cambridge Analytica whistleblower Christopher Wylie, Bannon realised that "to change politics you need to change culture. And fashion trends are a useful proxy for that. Trump is like a pair of Uggs, or Crocs, basically. So how do you get from people thinking 'Ugh. Totally ugly' to the moment when everyone is wearing them? That was the inflection point he was looking for".[5]

What this means is that politics and culture are now indistinguishable. This is not just a question of culture being explicitly political, or even of it reflecting the politics of its time, whether consciously or not. The claim is much stronger; in the world of social media there is no distinction between aesthetic forms of representation, and performative political enunciations. Politics, as a whole, becomes a massive work of art, albeit one that may not give a great deal of pleasure. This is surely something that Donald Trump fully understands, even if he would not necessarily articulate it in this way. This is precisely the rendering of politics as aesthetics, as performed by the Fascists in the 1930s, and written about by Walter Benjamin at that time.

There are many different historical moments that mark the beginning of our current predicament. In her book *Kill All Normies* (2017), Alison Nagle invokes the Marquis de Sade.[6] I tend to go back to the second half of the nineteenth century, and, in particular, the 1870s, after the publication of Karl Marx's *Das Kapital* in 1867 and before Friedrich Nietzsche's first declaration of the death of God in 1883. In 1877, the young American artist James McNeill Whistler sued the art critic John Ruskin for libel in a notorious trial, an event that, in retrospect can be seen as symbolising a momentous cultural shift. Twenty-five years earlier, in 1852, Ruskin had published the second volume of *The Stones of Venice,* which contained the chapter 'The Nature of Gothic' that inspired William Morris and Edward Burne-Jones to found the Arts and Crafts movement. In this much anthologised piece, Ruskin attempted to sketch out an ethics of art, using the example of the Gothic style as found in Venice.

It was while visiting the Grosvenor Gallery to see an exhibition of work by Burne-Jones that Ruskin came across a painting by Whistler entitled *Nocturne in Black and Gold: The Falling Rocket* (c.1875). Incensed by what he saw he wrote a coruscating notice in *Fors Clavigera,* his letters to the workmen of England:

> *For Mr. Whistler's own sake, no less than for the protection of the purchaser, Sir Coutts Lindsay ought not to have admitted works into the gallery in which the ill-educated conceit of the artist so nearly approached the aspect of wilful imposture. I have seen, and heard, much of Cockney impudence before now; but never expected to hear a coxcomb ask two hundred guineas for flinging a pot of paint in the public's face.*[7]

Whistler sued Ruskin for libel, which became one of the most notorious court cases of the time. The jury decided in Whistler's favour, but awarded him a farthing in damages, which led to his bankruptcy. This wasn't just about personal reputation, but was a battle for the soul of art, with Ruskin defending its social responsibilities and value, and Whistler believing in "art for art's sake". As Whistler famously declared in his book *The Gentle Art of Making Enemies,* written in response to the libel trial:

> *Art should be independent of all clap-trap — should stand alone, and appeal to the artistic sense of eye or ear, without confounding this with emotions entirely foreign to it, as devotion, pity, love, patriotism, and the like. All these have no kind of concern with it, and that is why I insist on calling my works "arrangements" and "harmonies".*[8]

For Ruskin, beauty was bound up with ethics and divinely ordained order, whereas for Whistler and those adhering to art for art's sake, there was no such connection. In a sense, the trial was more about this than about any libel. In some senses, the trial can be understood as the beginnings of artistic Modernism. It can also be seen as the point at which art bifurcated, producing two different paths: that which led from aestheticism and art for art's sake to avant-garde autonomy, and the other, which cleaved to the Ruskinian path of social engagement.

In the early 1870s, Ruskin had been appointed Slade Professor of Fine Art at Oxford University. In 1874, his disgust at the intellectual complacency of the students led him to initiate a project in which a number of undergraduates, including Oscar Wilde and Arnold Toynbee, built a road from the village of North Hinksey to South Hinksey across the swamp between the two. Toynbee went on to found Toynbee Hall in the East End of London, where Ashbee worked at one time, and in doing so, continued the Ruskinian legacy. Wilde, however, took the other path, the one that led to Aestheticism.

As many commentators have pointed out, Whistler and Wilde are the direct precursors of later figures such as David Bowie, with what Shelton Waldrep describes as their "aesthetics of self-invention".[9] In his book on Glam Rock, Simon Reynolds goes so far as to describe Wilde as the "first philosopher of glam, expounding its tenets eighty years in advance".[10] He suggests that Bowie's career is predicted "with Wilde's rhetorical question: 'Is insincerity such a terrible thing? I think not. It is merely a method by which we can multiply our personalities'".[11] For Reynolds, Wilde is the "prophet of glam" because of the "irrationalism that bubbles to the surface like intoxicating fumes" in his writing. Bowie's interest in masks, personae and mime seem perfectly consonant with Wilde's own attack on the "prison-house of realism". "Art should be a veil rather than a mirror", Wilde declares, and we "must cultivate the lost art of lying".[12]

Wilde was greatly influenced by Oxford don and critic Walter Pater. In 1873, Pater published *The Renaissance.* Though under-appreciated, Pater is now probably one of the most important figures in nineteenth-century art, especially in relation to the rise of what became known as Aestheticism. Indeed, *The Renaissance* could be thought of as the founding text of Aestheticism. Though Pater himself was also greatly influenced by

Ruskin, *The Renaissance* has been understood as a rebuttal and, even, a kind of negation of *The Stones of Venice.* Pater declared the need to "see the object as in itself it really is", which is really "to know one's own impression as it really is, to discriminate it, to realise it distinctly".[13] Therefore, "he who experiences these impressions strongly, and drives directly at the discrimination and analysis of them, has no need to trouble himself with the abstract question of what beauty is in itself, or what its exact relation to truth or experience is — metaphysical questions, as unprofitable as metaphysical questions elsewhere".[14]

Pater considered the brief, ephemeral sensation to be the most important experience in life, such as "a sudden light transfigures some trivial thing, a weather-vane, a wind-mill, a winnowing fan, the dust in the barn door. A moment — and the thing has vanished, because it was pure effect; but it leaves a relish behind it, a longing that the accident may happen again".[15] As he put it, "To burn always with this hard, gemlike flame, to maintain this ecstasy, is success in life". He famously declared what was clearly his manifesto for a life of pure sensation: that "all art constantly aspires towards the condition of music".[16]

Though *The Renaissance* was much admired by some, it was also condemned by others as decadent and endorsing amorality and hedonism. Pater's own homosexuality and his entanglements with Oxford undergraduates such as William Money Hardinge, known as "the Balliol Bugger", did not help his wider reputation. Despite his current obscurity, Denis Donoghue claims that Pater "is audible in virtually every attentive modern writer — in Hopkins, Wilde, James, Yeats, Pound, Ford, Woolf, Joyce, Eliot, Aiken, Hart Crane, Fitzgerald, Forster, Borges, Stevens", or, in other words, the entire literary modernism movement.[17]

Though Pater's concern may seem distant from our technological concerns, his writing takes place in the first modern media age, in which photography, telegraphy, the typewriter, and even the earliest computer had already been invented, to be joined very shortly by the gramophone, wireless, cinema, and even the beginnings of television technology. Walter Benjamin remarked on this in his essay "The Work of Art in the Age of its Technical Reproducibility" (1935):

> *For when, with the advent of the first truly revolutionary meansofproduction,namelyphoto-graphy,whichemerged*

> *at the same time as socialism, art felt the approach of the crisis which a century later has become unmistakable, it reacted with the doctrine of 'l'art pour l'art', that is, with a theology of art. This in turn gave rise to a negative theology, in the form of an idea of "pure" art, which rejects not only any social function of art, but any definition in terms of representational content.*[18]

Even though the Aestheticism he fostered was a reaction against new media such as photography, in a sense, Pater is also the first media theorist *avant la lettre.* As Sean Cubitt points out, in Pater's idea that all art aspires to the condition of music we find the beginnings of our modern understanding of how we engage with media. Cubitt suggests that "Pater was of the first generation to shut their eyes during a performance, since it is the purity of music, its distance from denotative meanings, its abstraction, to which the other media... aspired". Thus, the "purity of sound had to be conceptualised before it could be invented as technology".[19] For Cubitt, therefore:

> *The "purity" of music is the condition of its becoming a commodity: before it can be exchanged and consumed under the aegis of capital, music must first be divorced from the moment of its production. The labour of performance has to be subsumed into the autonomous status of an object before it can be understood as product, standing free of the material practice of its production, both steps essential to its circulation as commodity, organised into the forms of communication and profit-making which the social formation of capital demands.*[20]

But Benjamin sees even the most radical forms of art as prefiguring new media, especially film:

> *It has always been one of the primary tasks of art to create a demand whose hour of full satisfaction has not yet come. The history of every art form has critical periods in which the particular form strains aftereffects which can be easily achieved only with a changed technical standard — that is to say, in a new art form. The excesses and crudities of art which thus result, particularly in periods of so-called*

> *decadence, actually emerge from the core of its richest historical energies. In recent years, Dadaism has amused itself with such barbarisms. Only now is its impulse recognizable: Dadaism attempted to produce with the means of painting (or literature) the effects which the public today seeks in film.*

In his book *The Laws of Cool* (2004), Alan Liu refers to "the modernist graphic design movements that followed up on the avant-garde experiments of futurism and dadaism", such as Russian constructivism, De Stijl, the Bauhaus, and the New Typography. Designers working within these movements became less "artists of meaning than technicians of information", aiming for "clean, efficient information for an age drowning in media".[22] Clarity of information in design was paralleled by the "quest for purity" in radio signals, and, one might add, in one of the founding gestures of the post-war cybernetic culture, Claude Shannon's separation of the concept of information from the semantic content of any message — a kind of abstraction that we might characterise as "information for information's sake". In his essay "Avant Garde as Software", Lev Manovich states that:

> *Thus, new media does represent a new stage of the avant-garde. The techniques invented by the 1920s Left artists became embedded in the commands and interface metaphors of computer software. In short, the avant-garde vision became materialized in a computer. All the strategies developed to awaken audiences from a dream-existence of bourgeois society (constructivist design, New Typography, avant-garde cinematography and film editing, photomontage, etc.) now define the basic routine of a post-industrial society: the interaction with a computer.*[23]

Manovich ends his essay by noting that:

> *In short, the avant-garde becomes software. This statement should be understood in two ways. On the one hand, software codifies and naturalizes the techniques of the old avant-garde. On the other hand, software's new techniques of working with media represent the new avant-garde of the meta-media society.*[24]

If the avant-garde prefigures and, eventually, becomes software, then that software contains the legacy of the avant-garde's origins in Aestheticism. This is named and described by Liu as "cool". In *The Laws of Cool* (2004), Liu seeks to define this ineffable quality and map its emergence and its contradictions. For Liu, "cool" is a response to the fact that "knowledge work has no recreational outside". It is an "intraculture" rather than a subculture, a declaration that we may work in the grey cubicles of IT companies, but "we're cool".[25] He quotes a poem from the *Project Cool Website* from 1997, which offers its definition of cool:

> There is no one definition of cool.
> There is no one definition of beauty.
> Art
> Obscenity
>
> It's a sort of
> "I know it when I see it"
> type of thing. You can argue
> 'til the cows come home
> that this was or wasn't cool,
> but it's all pretty subjective.[26]

For Liu, "cool" is the style of information; the style that arises out of information work. It is both a response to, and a reaction against the instrumentality of information culture. Rather than aiding the transparent transmission of information, "cool" makes it mysterious. Liu suggests that the "koan of cool can be put as follows: *we know what is cool, but part of what we know is that we cannot know what we know. Cool forbids it"*.[27]

Later, he maintains that "Cool is the code [...] for awareness of the information interface". We don't stare through the window of the screen to information, but at the screen itself, much as we look at "the gorgeousness of stained-glass windows themselves" in Gothic cathedrals. Liu invokes Coleridge's idea of the "translucence of symbol".[28] However, Liu claims that cool does not have "anything to do *per se* with the utility of information". Rather, it is an "ethos against information". He quotes the Critical Art Ensemble quoting Georges Bataille, to the effect that "the end of technological progress be neither apocalypse nor utopia, but simply uselessness".[29]

Liu argues that the kind of information design that engendered the ethos of cool can best be understood through formalism — whether that of the New Criticism, or Russian Formalism:

> *Formalism, above all other twentieth-century artistic and critical movements, suborned the technological rationality of modernity by remolding its functionalist assumptions so profoundly as to imprint them with the distinctive style of "modernism".*[30]

But, as Liu points out, if the cool design on and off computers is a legacy of modernist design, it also produces its antithesis: anti-design. Liu cites the various experiments in disruptive design such as Californian New Wave, Deconstructionist design, Memphis, New Wave Typography, the work of David Carson, and that coming out of the Cranbrook School of Art in the 1980s and 1990s. Disruptive design took advantage of programs such as Photoshop to critique, deconstruct, and undermine notions of clarity and the frictionless communication of information.[31]

One of Liu's most important points is that cool is "among the most totalitarian aesthetics ever created".[32] It is low in affect, lacking the means to express the more tragic, horrifying, beautiful or sublime elements of experience. "Before all the horrors and despairs offered up on even ordinary journalism Web sites, cool is wordless, or at best responds 'That's uncool.'" However Liu points out that, "All terror, anger, lust, joy and so on thus bleed out of cool to manifest with compensatory, even artificial, fervor in personal e-mail, alt. newsgroups, chat, hate sites, porn, and other parts of the Internet that sequester themselves from post-industrial knowledge work by being intractably 'unproductive'".[33] What Liu could not have known at the time of his book's publication is the degree to which those elements of the Web would become the dominant modes of discourse in our contemporary digital culture, with its dark fascistic overtones.

This brings me back to Bowie. In 1977, he released his album *"Heroes"* (note the ironic quotation marks). The cover is a black and white photograph of Bowie in a contorted pose based, apparently, on the work of Austrian artist Egon Schiele. The cover, in particular, is an almost ideal representation of a certain ethos, something glamorously grey and European, and bound up with the Second World War and the Cold War. It is also astonishingly cool, channelling the ethos of the twentieth century avant-garde. In Agata Pyzik's *Poor but sexy: Culture clashes in Europe East and West (*2014), the author tries to recover a sense of the dreamworld of Cold War Eastern Europe and analyse how it became an object of fascination for westerners such as Bowie. Pyzik has coined the useful term "Berlinism" to capture this elusive set of phenomena:

> *What I'll be calling "Berlinism" is the twentieth-century phenomenon of the German capital as a dreamland for both easterners and westerners. Arguably, it starts after the First World War, when the Weimar era turns it into a capital of all sorts of debauchery and transgression, in culture, politics, literature, art, music and theatre.*[34]

It was Bowie who took all the elements of German and Eastern European pre- and post-war culture, and "put them all together". Bowie, a "model postmodernist, someone who built his life and art out of the artificial, the fabricated, who went through pop art, comic books and Brecht, needed the necessary frisson of the real, which he found in Berlin, Warszawa and Moscow".[35] This frisson he bequeathed to punk and postpunk bands such as Joy Division, and to New Romantics such as Ultravox, and Visage. They all sought that strange negative glamour of the dreamworld of capitalism's other in the later stages of the Cold War. Bowie was "obsessed with certain elements of modernity. He was driven to German culture, especially the Weimar period, expressionism, *Neue Sachlichkeit,* theatre, Brecht".[36] Part of this fascination, unfortunately, took the form of an advocacy of Fascism. The records he made while living in Berlin, particularly *"Heroes"* and *Low,* "are psychogeographical albums, where he takes us on various trips to places charged with history, various stops around Berlin, Neukölln, the Wall; then Warszawa, Japan, China, yearning for the East".[37]

"Heroes" is a more subtle version of the endless fascination with twentieth-century Germany in 1970s' British pop culture. Punks sported Nazi imagery such as Swastikas. Bands had names that alluded to aspects of the War and the Holocaust, such as Joy Division, and its successor New Order. Writing about the film *Downfall* (2004), about the fall of Nazi Germany, Mark Fisher alludes to the Punk use of Nazi imagery:

> *While those scenes play out, you can almost hear Johnny Rotten leering, "when there's no future how can there be sin?" [...] It's no accident that post-punk in many ways begins here. As the Pistols pursue their own line of abolition into the scorched earth nihilism of "Belsen was a Gas" and "Holidays in the Sun", they keep returning to the barbed-wire scarred Boschscape of Nazi Berlin and the Pynchon Zone it became after the war. Siouxsie famously sported a swastika for a while, and although much of the flaunting of the Nazi imagery was supposedly for superficial shock effects, the punk-Nazi connection was about much more than trite transgressivism. Punk's very 1970s, very British fixation on Nazism posed ethical questions so troubling they could barely be articulated explicitly: what were the limits of liberal tolerance? Could Britain be so sure that it had differentiated itself from Nazism (a particularly pressing issue at a time that the NF was gathering an unprecedented degree of support)? And, most unsettling of all, what is it that separates Nazi Evil from heroic Good?*[38]

More generally, Germanness held a certain fascination, as if it contained the secret to the emerging world of new technologies and architectures. Germany's own rock music, *cosmische Musik,* known, rather vulgarly in Britain as Krautrock, influenced New Wave here and in the States, not least in terms of style. Kraftwerk, in particular, epitomised — parodically, perhaps — the idea of German machine-like efficiency and modernity. Bowie, ever alert to what was new and of the moment, spent a number of years in Berlin with Iggy Pop, producing some of his best work there, as well as that of Pop and Lou Reed.

1977 is also the year when Steve Wozniak and Steve Jobs founded Apple. For many, the products of Apple Computers are the epitome of digital cool. Their look, from the Macintosh onwards, exemplifies the clean lines and purity of modernist design. Apple owed much of its design ethos to the German tradition of industrial design, going back to Peter Behrens's work with AEG in the early-twentieth century, and on to Dieter Rams's involvement with companies such as Braun. Rams's designs are the acme of the aesthetic of minimalism, offering, as the name of a book on his work has it, *As Little Design as Possible.*[39]

For Franco 'Bifo' Berardi, 1977 is the crucial year in which everything changes. Berardi claims that it is not just the year when "Steve Wozniak and Steven Jobs created the trademark of Apple and, what is more, created the tools for spreading information technology",[40] but also when Alain Minc and Simon Nora wrote *The Computerization of Society,* when Jean François Lyotard wrote the book *The Postmodern Condition,* and when Charlie Chaplin died. "This was the year of the end of the twentieth century: the turning point of modernity". In his more recent book, *Heroes* (2015), named in part after Bowie's album, Berardi shows how 1977 is the year that leads to the epidemic of meaningless mass murder and suicide that characterises an era obsessed by relentless competition and hyperconnectivity.

The reason why *"Heroes"* is so important, along with a cluster of albums that came out at the same time, is that it and they represent the moment when the new world we now live in was born out of the ruins of the older post-war world. As Hito Steyerl argues, Bowie represents the emergence of a brand-new hero, suitable for, and just in time for the neoliberal revolution that is around the corner:

> *Bowie's hero is no longer a subject, but an object: a thing, an image, a splendid fetish—a commodity soaked with desire, resurrected from beyond the squalor of its own demise. Just look at a 1977 video of the song to see why: the clip shows Bowie singing to himself from three simultaneous angles, with layering techniques tripling his image; not only has Bowie's hero been cloned, he has above all become an image that can be reproduced, multiplied, and copied, a riff that travels effortlessly through commercials for almost anything, a fetish that packages Bowie's glamorous and unfazed post-gender look as product. Bowie's hero is no longer a larger-than-life human being carrying out exemplary and sensational exploits, and he is not even an icon, but a shiny product endowed with post-human beauty: an image and nothing but an image. This hero's immortality no longer originates in the strength to survive all possible ordeals, but from its ability to be xeroxed, recycled, and reincarnated. Destruction will alter its form and appearance, yet its substance will be untouched. The immortality of the thing is its finitude, not its eternity.*[42]

In his later career and life, Bowie did an excellent job of appearing to be, as near as possible given the circumstances, a normal person, living quietly in New York with his wife and daughter. It is, perhaps, no coincidence that this period of normality was also his least creative, including as it did, the nadir of his career, *Tin Machine.* However, when he was doing his best work, he was also clearly a very disturbed and strange individual. As has been recounted many times, the period in which he made many of his greatest records was also the time when he was living on a diet of milk, peppers and cocaine, deeply involved with Crowleyan occult beliefs, exorcisms, the Kabbalah and suffering from delusions and paranoia. It was also in this period that he flirted most fiercely with right wing ideas. As Simon Reynolds puts it in his history of Glam Rock, even if Bowie claimed his cocaine abuse as mitigating circumstances, his remarks of this sort "don't read as addled provocations. They are so frequently, so articulately argued, so consistently *excited* in tone, it's hard to avoid concluding that Bowie had developed a morbid fascination with Fascism".[43] As far back as 1969, he had proclaimed in an interview with *Music Now!* magazine that Britain was "crying out for a leader" and named Enoch Powell as the best candidate.[44] In later interviews he compared himself, or at least Ziggy Stardust, to Hitler. In 1976, in *Rolling Stone,* interviewed by Cameron Crowe, he proclaimed that:

> *I fell for Ziggy too. It was quite easy to become obsessed night and day with the character. I 'became' Ziggy Stardust. David Bowie went totally out the window. Everybody was convincing me that I was a Messiah, especially on that first American tour. I got hopelessly lost in the fantasy. I could have been Hitler in England. Wouldn't have been hard. Concerts alone got so enormously frightening that even the papers were saying, 'This ain't rock music, this is bloody Hitler! Something must be done!' And they were right. It was awesome. Actually, I wonder [...] I think I might have been a bloody good Hitler. I'd be an excellent dictator. Very eccentric and quite mad.*[45]

He goes on to muse that he "should be prime minister of England", adding: "I wouldn't mind being the first English president of the United States either. I'm certainly right wing enough". In the same year, in an interview in *Playboy* magazine, again with Cameron Crowe, he is asked about this suggestion: "You've often said that you believe very strongly in fascism. Yet, you also claim you'll one day run for Prime Minister of England. More media manipulation?" To which Bowie replies:

> *I'd love to enter politics. I will one day. I'd adore to be Prime Minister. And, yes, I believe very strongly in fascism. The only way we can speed up the sort of liberalism that's hanging foul in the air at the moment is to speed up the progress of a right-wing, totally dictatorial tyranny and get it over with as fast as possible. People have always responded with greater efficiency under a regimental leadership. A liberal wastes time saying, "Well, now, what ideas have you got?" Show them what to do, for God's sake. If you don't, nothing will get done. I can't stand people just hanging about. Television is the most successful fascist, needless to say. Rock stars are fascists, too. Adolf Hitler was one of the first rock stars.*[46]

Also, in 1976, Bowie informed a Swedish newspaper that "Britain could benefit from a fascist leader",[47] though he clarified this by explaining that he meant fascism in its true sense. It was in the same year that he returned to England on the Orient Express, arriving at Victoria Station to be met by an open-top Mercedes, a favoured form of transport for the Nazis. It was then that he may or may not have made a Nazi salute to the crowd. There is, unfortunately, plenty of other evidence in the dossier marked "David Bowie's Fascist Tendencies".

Perhaps, the strangest thing about reading these proclamations by Bowie is how much they remind me of Donald Trump. In a sense, Trump is both the mirror of Bowie as much as Bowie is the missing link between Aestheticism, with Whistler, Wilde, and Pater in the late-nineteenth century. Interestingly, Trump and Bowie were of the same generation, with the former born in June 1946, and the latter in January 1947, a mere eight months apart. They even both hung out at Studio 54 in New York in the 1970s and 80s, though I do not know if they were ever there at the same time. Trump and his then wife Ivana went to the opening. Whatever the truth of Bowie's politics, or Trump's for that matter, both understood something profound about our contemporary culture: that everything is image. What they did with that understanding is where they differ. Bowie produced some of the greatest music of the twentieth century. In the twenty-first century, Trump has become, unfortunately, the most powerful man in the world.

Perhaps Trump's "Mini-Me", Boris Johnson, understands this connection better than it might appear. For his final speech as London Mayor Johnson invoked the recently deceased Bowie as being all that the terrorist group ISIS abhorred.

> *Think of that great Londoner who died this week amid an unexpected outpouring of global grief, a man who was born as Davy Jones from Brixton — and yet who reinvented himself as Ziggy Stardust and the Thin White Duke and other characters and who was recognised as a genius. It is hard to think of anything that would be more repugnant to the morons of ISIL than the Bowie phenomenon. And yet it is that willingness to encourage individualism, and eccentricity, and experiment, that is one of the main drivers of the genius of modern London. Where else would you find ginger-bearded hipsters selling Froot Loops for £3.80 per bowl? Where else but London would you find a restaurant where you are served in total darkness by blind waiters, the contention being that you will somehow taste your food better? Where else would you find a cocktail bar in a public toilet?*

It's hard to imagine a statement that reveals more of the emptiness of contemporary British culture, and also more about how Johnson sees himself. As with his biography of Churchill, his description of Bowie as a genius of reinvention is really a kind of self-portrait. The examples of London's "individualism, and eccentricity, and experiment" are the most banal and irrelevant forms of aestheticised and commodified "cool". Far worse is the fact that the "ginger-bearded hipsters" are those whose presence in places such as Hoxton are destroying the older communities in those areas. They are the vanguard of neoliberal creative destruction. Johnson, in ways he may not appreciate, is a child of Bowie and of the latent and sometimes blatant Fascism in Bowie's aestheticism. Bowie, in turn, is a descendent of Wilde, Whistler, and Pater, the great English advocates of Aestheticism and art for art's sake. This Aestheticism, in its influence on twentieth century avant-garde art and design, is also the underlying ethos of "cool" in contemporary networked culture, and its complementary uncool element of hate, trolling and violence.

For Benjamin, the ultimate result of art for art's sake is the aestheticisation of war as proclaimed and celebrated by Fascism. Benjamin, quotes Marinetti's Futurist Manifesto:

> *For twenty-seven years we Futurists have rebelled against the branding of war as anti-aesthetic [...] Accordingly we state [...] War is beautiful because it establishes man's dominion over the subjugated machinery by means of gas masks, terrifying mega-phones, flame throwers, and small tanks. War is beau-tiful because it initiates the dreamt-of metallization of the human body. War is beautiful because it enriches a flowering meadow with the fiery orchids of machine guns. War is beautiful because it combines the gunfire, the cannonades, the cease-fire, the scents, and the stench of putre-faction into a symphony. War is beautiful because it creates new architecture, like that of the big tanks, the geometrical formation flights, the smoke spirals from burning villages, and many others. [...] Poets and artists of 20 Futurism! [...] remember these principles of an aesthetics of war so that your struggle for a new literature and a new graphic art [...] may be illumined by them!"*[49]

Benjamin ends the "Work of Art in the Age of Mechanical Reproduction" with his famous last paragraph — a call to arms to combat the aestheticising of politics with the politicising of aesthetics:

> *"Fiat ars-pereat mundus", says fascism, expecting from war, as Marinetti admits, the artistic gratification of a sense perception altered by technology. This is evidently the consummation of 'l'art pour l'art'. Humankind, which once, in Homer, was an object of contemplation for the Olympian gods, has now become one for itself. Its self-alienation has reached the point where it can experience its own annihilation as a supreme aesthetic pleasure. Such is the aestheticizing of politics, as practiced by fascism. Communism replies by politicizing art.*[50]

Notes

1. Virginia Woolf, *Mr Bennett and Mr Brown* (London: Hogarth Press, 1924), 4.

2. Charles Jencks, *The Language of Postmodern Architecture* (New York: Rizzoli, 1977), 9.

3. "David Bowie, Interview with Jeremy Paxman", *BBC Newsnight,* BBC Two, 3 December 1999, https://www.youtube.com/watch?v=FiK7s_OtGsg.

4. Caroline Cadwalladr, "'I made Steve Bannon's psychological warfare tool': meet the data war whistle-blower", *The Guardian,* 17 March 2018, https://www.theguardian.com/news/2018/mar/17/data-war-whistle-blower-christopher-wylie-faceook-nix-bannon-trump.

5. Cadwalladr, "'I made Steve Bannon's psychological warfare tool".

6. Alison Nagle, *Kill All Normies: Online culture wars from 4chan and Tumblr to Trump and the alt-right* (Winchester, UK: Zer0 Books, 2017), 28.

7. John Ruskin, et al. *The Works of John Ruskin,* vol. VII, ed. E. T. Cook and Alexander Wedderburn, Library edn. (London: George Allen, 1903), 160.

8. James McNeill Whistler, *The Gentle Art of Making Enemies* (London: William Heinemann, 1890), 35.

9. Shelton Waldrep, *The Aesthetics of Self-Invention: Oscar Wilde to David Bowie* (Minneapolis: University of Minnesota Press, 2004).

10. Simon Reynolds, *Shock and Awe: Glam Rock and its Legacy, from the Seventies to the Twenty-First Century* (London: Faber & Faber, 2016), 91.

11. Reynolds, *Shock and Awe,* 91.

12. Reynolds, *Shock and Awe,* 92

13. Walter Pater, *The Renaissance* (London: Macmillan & Co., 1910), viii.

14. Pater, *The Renaissance,* viii.

15. Pater, *The Renaissance,* 176.

16. Pater, *The Renaissance,* 135.

17. Denis Donoghue, *Walter Pater: Lover of Strange Souls* (New York: Knopf, 1995).

18. Walter Benjamin, *Selected Writings, 1935–1938,* vol. 3, ed. Marcus Bullock and Michael W. Jennings (Cambridge, MA: Belknap Press, 2002) 105–106.

19. Sean Cubitt, *Timeshift: On Video Culture* (Oxford: Routledge, 1991), 44.

20. Cubitt, *Timeshift,* 44.

21. Benjamin, *Selected Writings,* 118.

22. Alan Liu, *The Laws of Cool: Knowledge Work and the Cultures of Information* (Chicago: University of Chicago Press, 2004), 199.

23. Lev Manovich, "Avant-Garde as Software", 2002, http://manovich.net/content/04-projects/028-avant-garde-as-software/24_article_1999.pdf.

24. Manovich, "Avant-Garde as Software".

25. Liu, *The Laws of Cool,* 77–8.

26. Liu, *The Laws of Cool,* 177.

27. Liu, *The Laws of Cool,* 179.

28. Liu, *The Laws of Cool,* 183.

29. Liu, *The Laws of Cool,* 185–6.

30. Liu, *The Laws of Cool,* 195–6.

31. Liu, *The Laws of Cool,* 216.

32. Liu, *The Laws of Cool,* 237.

33. Liu, *The Laws of Cool,* 237–8.

34. Agata Pyzik, *Poor but sexy: Culture clashes in Europe East and West* (Winchester, UK: Zer0 Books, 2014), 78–9.

35. Pyzik, *Poor but sexy,* 105.

36. Pyzik, *Poor but sexy,* 81.

37. Pyzik, *Poor but sexy,* 87.

38. Mark Fisher, *K-Punk: The Collected and Unpublished Writings of Mark Fisher* (London: Repeater, 2018), 132.

39. Sophie Lovell, *Dieter Rams: as little design as possible* (London: Phaidon, 2011).

40. Franco 'Bifo' Berardi, *Precarious Rhapsody: Semiocapitalism and the pathologies of the post-alpha generation* (London: Minor Compositions, 2009), 14.

41. Berardi, *Precarious Rhapsody,* 14–15.

42. Hito Steyerl, *The Wretched of the Screen* (Berlin: Sternberg Press, 2012), 49.

43. Reynolds, *Shock and Awe,* 548.

44. Kate Simpson, "David Bowie Interview", *Music Now!,* 20 December 1969.

45. Cameron Crowe, "David Bowie: Ground Control to Davy Jones", *Rolling Stone Magazine,* 12 February 1976, https://www.rollingstone.com/music/music-news/david-bowie-ground-control-to-davy-jones-77059/.

46. Cameron Crowe, "Playboy Interview: David Bowie", *Playboy Magazine,* September 1976, http://www.theuncool.com/journalism/david-bowie-playboy-magazine/.

47. David Bowie in Tony Stewart, "Heil And Farewell", *New Musical Express,* 8 May 1976, 9.

48. Matt Dathan, "The Spirit of David Bowie can beat Isis says Boris Johnson", *The Independent,* 15 January 2016, https://www.independent.co.uk/news/uk/politics/the-spirit-of-david-bowie-can-beat-isis-says-boris-johnson-a6813866.html.

49. Benjamin, *Selected Writings,* 121.

50. Benjamin, *Selected Writings,* 122.

References

Benjamin, Walter. *Selected Writings, 1935–1938,* vol. 3, edited by Marcus Bullock and Michael W. Jennings. Cambridge, Mass.: Belknap Press, 2002.

Berard, Franco 'Bifo'. *Precarious Rhapsody: Semiocapitalism and the pathologiesofthepost-alphageneration.* London: Minor Compositions, 2009.

Cadwalladr, Caroline. "'I made Steve Bannon's psychological warfare tool': meet the data war whistleblower.' *The Guardian,* 17 March 2018. https://www.theguardian.com/news/2018/mar/17/data-war-whistleblower-christopher-wylie-faceook-nix-bannon-trump.

Crowe, Cameron. "David Bowie: Ground Control to Davy Jones". *Rolling Stone Magazine,* 12 February 1976. https://www.rollingstone.com/music/music-news/david-bowie-ground-control-to-davy-jones-77059.

Crowe, Cameron. "Playboy Interview: David Bowie". *Playboy Magazine.* September 1976. http://www.theuncool.com/journalism/david-bowie-playboy-magazine.

Cubitt, Sean. Timeshift: *On Video Culture.* New York: Routledge, 1991.

Dathan, Matt. "The Spirit of David Bowie can beat Isis says Boris Johnson". *The Independent,* 15 January 2016. https://www.independent.co.uk/news/uk/politics/the-spirit-of-david-bowie-can-beat-isis-says-boris-johnson-a6813866.html.

Donoghue, Denis. *Walter Pater: Lover of Strange Souls.* New York: Knopf, 1995.

Fisher, Mark. *K-Punk: The Collected and Unpublished Writings of Mark Fisher.* London: Repeater, 2018.

Jencks, Charles. *The Language of Postmodern Architecture.* New York: Rizzoli, 1977.

Liu, Alan. *The Laws of Cool: Knowledge Workandthe CulturesofInformation.* Chicago: University of Chicago Press, 2004.

Lovell, Sophie. *Dieter Rams: as little design as possible.* London: Phaidon, 2011.

Manovich, Lev. "Avant-Garde as Software". 2002. http://manovich.net/content/04-projects/028-avant-garde-as-software/24_article_1999.pdf.

Nagle, Alison. *Kill All Normies: Online culture wars from 4chan and Tumblr to Trump and the alt-right.* Winchester: Zer0 Books, 2017.

Pater, Walter. *The Renaissance.* London: Macmillan & Co., 1910.

Pyzik, Agata. *Poor but sexy: Culture clashes in Europe East and West.* Winchester, UK: Zer0 Books, 2014.

Reynolds, Simon. *Shock and Awe: Glam Rock and its Legacy, from the Seventies to the Twenty-First Century.* London: Faber & Faber, 2016.

Ruskin, John, Cook, E. T. and Wedderburn, Alexander. *The Works of John Ruskin,* vol. VII, edited by E. T. Cook and Alexander Wedderburn, Library edn. London: George Allen, 1903.

Stewart, Tony. "Heil And Farewell". *New Musical Express,* 8 May 1976, 9.

Steyerl, Hito. *The Wretched of the Screen.* Berlin: Sternberg Press, 2012.

Waldrep, Shelton. *The Aesthetics of Self-Invention: Oscar Wilde to David Bowie.* Minneapolis: University of Minnesota Press, 2004.

Whistler, James McNeill. *The Gentle Art of Making Enemies.* London: William Heinemann, 1890.

Woolf, Virginia. *Mr Bennett and Mr Brown.* London: Hogarth Press, 1924.

Explanatory Publics: Explainability and Democratic Thought

David M. Berry

In order to legitimate and defend democratic politics under conditions of computational capital, my aim is to contribute a notion of what I am calling *explanatory publics.* I will explore what is at stake when we question the social and political effects of the disruptive technologies, networks and values that are hidden within the "black boxes" of computational systems. By "explanatory publics", I am gesturing to the need for frameworks of knowledge — whether social, political, technical, economic, or cultural — to be justified through a *social right to explanation.* That is, for a polity to be considered democratic, it must ensure that its citizens are able to develop a capacity for explanatory thought (in addition to other capacities), and, thereby, able to question ideas, practices, and institutions in society. This is to extend the notion of a public sphere where citizens are able to question ideas, practices, and institutions in society more generally.[1] But it also adds the corollary that citizens can demand explanatory accounts from institutions and, crucially, the digital technologies that they use. I agree with Outhwaite that "what makes an explanation, or what makes an explanation a good one, is therefore a difficult question, which may require a detailed study, not just of the logical properties of the explanation but of the context in which it was offered".[2]

This is important in computational capitalism because when we call for an *explanation,* we are able to understand the contradictions within this historically specific form of computation that emerges in late capitalism. These contradictions are continually suppressed in computational societies, but generate systemic problems borne of the need for the political economy of software to be obscured, so that its functions and the mechanisms of value generation are hidden from public knowledge. Why should this fundamental computational political economy be concealed? One reason is that an information society requires a form of public justification in order to legitimate it as an accumulation regime and to maintain trust. Trust is a fundamental basis of any system and has to be stabilised through the generation of norms and practices that create justifications for the way things are. This is required, in part, because computation is increasingly a central aspect of a nation's economy, real or imaginary. The suppression of its political economy is required, because computation rapidly destabilises the moral economy of capitalism, creating vast profits from exchange and production processes that might be considered pre-capitalistic or obscenely inegalitarian such as intensive micro-work or fragmented labour in the gig economy. These more recent experiments with micro-task production are nothing less than attempts to reinvent the world as a post-factory society. It requires the building of a new infrastructure of production by enclosing labour-power within algorithmic "wrappers" that present the surface effect of a seemingly unending stream of abstract labour. This labour-power is made available via websites and apps, creating a highly alienated form of labour-power that is disciplined and managed algorithmically through various forms of "signal" mechanisms that are generated by the system such as pay, ratings, reviews, and metrics. The "boss" of the old factory is abolished by computation and replaced by the algorithm that guides, chides, and informs through a personal device such as a smartphone, whose very intimacy makes it compelling and trustworthy.

Further, many of the sectors affected by computation are increasingly predicated on the illegitimate manipulation or monopolisation of markets, or are heavily data extractive. These effects threaten individual liberty, undermining a sense of individual autonomy, and even that bulwark of the neoliberal system: consumer sovereignty. Profit from computation also often appears to require the mobilisation of persuasive technologies that cynically, but very successfully, manipulate addictive human behaviour. Therefore, one of the key questions we need to ask is: How much computation can a society withstand? We can only answer this question if we create new forms of explanatory publics that have competences to discuss, critique, and challenge computational technical systems.

Google, at least at one point, internally understood the problem of an excess of computational power in terms of what it called a "creepy line".[3] Within the line, public acceptance of computation generates huge profits (or "good computation"), and outside of which computation is able to create effects that would be politically, or economically problematic, or even socially destructive, but which might generate even larger profits (or "bad computation"). The founders of Google, Larry Page and Sergey Brin, gestured towards this in their famous paper from 1998, "The Anatomy of a Large-Scale Hypertextual Web Search Engine", where they warned that if the "search engine were ever to leave the 'academic realm' and become a business, it would be corrupted. It would become 'a black art' and 'be advertising oriented'".[4] As Carr describes,

> *That's exactly what happened — not just to Google but to the internet as a whole. The white-robed wizards of Silicon Valley now ply the black arts of algorithmic witchcraft for power and money. They wanted most of all to be Gandalf, but they became Saruman.*[5]

Peter Thiel, a PayPal co-founder and chairman of Palantir, revealed a similar tendency made possible by computation when he identified the importance of software companies securing a technical monopoly. He termed this as a movement from "zero to one", the "one" representing the successful monopolisation of a technical niche or sector of the economy.[6] While this is not necessarily a surprise, the candour with which the Silicon Valley elite advocate for these economic structures, which are contrary to neoliberalism, let alone social democracy, should give us pause for thought. Indeed, Thiel goes so far as to argue that he "no longer believe[s] that freedom and democracy are compatible".[7] But while the profit-oriented organisation of a capitalist economy is unchanged, what is new is that exploitative processes function at a new intensity, and at all levels of society due to computation. It is no longer just the workers who are subject to processes of automation, but also the owners of capital themselves and, inevitably, their private lives. That the millionaires and billionaires of the technology industry should feel a need to protect their own families from the worst aspects of computation, with Steve Jobs famously withholding computers from his children and Larry Page, one of the co-founders of Google, managing to keep his personal life and even his children's name secret, is ironic given Google's mission "to organise the world's information and make it universally accessible and useful".[8] Unfortunately, this disconnectionism is not an option available to the majority of the world's population — even as it becomes a bourgeoise aspiration through digital detox camps and how-to-disconnect-guides in national newspapers.

The contradictions generated by this new system can be observed in discourse. Concepts carry over from the computational industries and spread as explanatory ideas across society. Indeed, we see principles from software engineering offered up for social engineering, with open source identified as an exemplar principle of organisation; platforms as future models for governance; calculation substituted for thought; and social media networks replacing community. This can also be seen when computation is described through dichotomies such as transparent and opaque, open and closed, augmentation and automation, freedom and subjugation, resistance and hegemonic power, the future of the economy, and its destruction.

If we focus on two of these discursive categories, augmentation and automation, we can see how they are used to orient and justify further computation. As far back as 1981, Steve Jobs, then CEO of Apple, famously called computers "Bicycles for the Mind", implying that they *augmented* the cognitive capacities of the user, making them faster, sharper, and more knowledgeable. He argued that when humans "created the bicycle, [they] created a tool that amplified an inherent ability [...] The Apple personal computer is a 21st century bicycle if you will, because it's a tool that can amplify a certain part of our inherent intelligence [...] [It] can distribute intelligence to where it's needed".[9] This vision has been extremely compelling for technologists and their apologists, who omitted to explain that these capacities might be reliant on wide-scale surveillance technology. But whilst this vision of bicycles for the mind might have been true in the 1980s, changes in the subsequent political economy of our societies means that computers are increasingly no longer augmenting our abilities, but rather *automating* them. Algorithms then become Weberian "iron cages", in which citizens are trapped and monitored by software, with code that executes faster than humans can think, overtaking their capacity for thought. Augmentation, which extends our capacity to do things, and automation, which replaces this capacity, show how analysing the historically specific examples of key dyadic concepts is crucial for understanding this struggle over the future of computation. Indeed, in response, we need a way of transcending these dichotomies, linking computation to a call for a social right to explanation — through what I call *explanatory* publics.

One of the ways in which we can do this is by recognising how information economies are founded on an attempt to make *thought subject to property rights.*[10] Principles of reasoning, mathematical calculation, logical operations, and formal principles necessarily become owned and controlled for an informational economy to function. But these forms of thought also become recast as the only legitimate forms of reason, feeding back into a new image of thought. Data is increasingly associated with wealth and power, linked explicitly with computational resources that submit this data to rapid computation and pattern-matching algorithms through machine-learning and related techniques. Humans can now purchase thinking capacity, whether through special algorithms or the augmentation possibilities of personal devices. Information processing is now so fast that it can be performed in the blink of an eye, and the results used to augment, if you can afford it, or else persuade, and potentially manipulate others who cannot. Depending on the price you're willing to pay, digital corporations can sell algorithms to either increase, or undermine an individual's reasoning capacities, and thereby supplement or substitute artificial analytic capacities that bypass the function of reason. For the wealthier, they have the option to literally buy better algorithms, better technologies, better capacities for thought. For the rest of us, algorithms overtake human cognitive faculties by shortcutting individual decisions by making a digital "suggestion" or "nudge". Indeed, many technology companies rely on techniques developed in casinos to nudge behaviour to maximise profitability such as creating addictive experiences and by disarming the will of the user.[11] As a consequence, cognitive inequality emerges in relation to a new neuro-diversity created by augmenting, or automating thought itself, potentially undermining democratic and public values. An example of how computation can differently segment the market is presented by the Amazon Kindle, which comes in two varieties: a cheaper version that contains constantly updating advertising on its home and lock screen ("With Special Offers"), and a more expensive version that is free of adverts ("Without Special Offers").

We might also note that the actually existing informational economy is built increasingly on software that has steered capitalism towards a data-intensive form of extractive economy — what Zuboff has termed "surveillance capitalism", and Stiegler has identified as "the automatic society".[12] This has been achieved through spying on users, data capture, arbitrage, and the manipulation of markets, but also, crucially, through facilitating monopolies of data — by using digital rights management, copyright, or patents. One of the scandals of contemporary capitalism is the extent to which widescale data capture and monitoring of users, their private lives, and their economic activities has been facilitated by computation. Not only has this been relatively unregulated, but it has allowed companies to assume that this kind of wholesale spying on people is the new normal and an acceptable practice in business. The scale of automated data accumulation is completely without precedent historically. Just taking Facebook as an example, we see almost continuous data collection on over 1.2 billion people worldwide. So much data, in fact, that even the CIA, the US intelligence agency, has signalled its inability to deal with the overload from what they call "digital breadcrumbs".[13] Indeed, Ira "Gus" Hunt, the CIA's former chief technology officer, has argued that "the value of any piece of information is only known when you can connect it with something else that arrives at a future point in time [...] Since you can't connect dots you don't have [...] we fundamentally try to collect everything and hang on to it forever".[14]

This has led to the idea that perhaps data contains highly lucrative insights that can create new sources of profit. While discourses about how "data is the new oil" have circulated, inevitably, those companies keen not to miss a profitable opportunity have put aside their caution and defaulted to maximum data collection whenever and wherever possible. Opting out of this surveillance regime has also become progressively more difficult, and, in my own case, attempting to "opt out" of the numerous data collection companies associated with the *Huffington Post,* for example, took three hours of frustrating clicking through numerous privacy statements. Even then, key links and options would be disguised as "hidden links", sliders that "were not available at the moment" and other techniques of dissuasion.[15] These companies build these systems deliberately in a user-hostile way, while keeping the default of data collection, even of minors and others who cannot legally give consent, in a system that, nonetheless, is structured in such a way as to unequally distribute the effects of data collection and algorithmic profiling to the poorer, less well-educated segments of society.

These data monopolies signal fundamental contradictions at the heart of computational capital, which appears to require legitimation through a façade of progress, individual choice and an enlightened technical philosophy, while the actually existing underlying political economy is increasingly structured around distortion, deception, and data capture.[16] These conditions create a form of cynical reason, particularly in the software industry, that I term *neo-computationalism.* This is an ideology subscribed to by an unhappy consciousness, which disavows the use of unsavoury data collection and surveillance techniques, while continuing to practice it. It is a system of thought that holds to a belief that social problems can be solved using more computation, at the same time as creating technical systems and algorithms that make them worse or amplify their pathologies. This contradiction at the level of both political economy and individual consciousness is destabilising to society and cannot be kept in check without the mobilisation of a set of justificatory discourses through the ideology of neo-computationalism.

Peter Sloterdijk has described cynical reason as enlightened false consciousness, which "is afflicted with the compulsion to put up with preestablished relations that it finds dubious, to accommodate itself to them, and finally even to carry out their business". Sloterdijk quotes Gottfried Benn, who explains modern cynicism as that which is lived as a private disposition that requires you,

> *to be intelligent and still perform one's work, that is unhappy consciousness in its modernized form, afflicted with enlightenment. Such consciousness cannot become dumb and trust again; innocence cannot be regained. It persists in its belief in the gravitational pull of the relations to which it is bound by its instinct for self-preservation. In for a penny, in for a pound. At two thousand marks net a month, counterenlightenment quietly begins; it banks on the fact that all those who have something to lose come to terms privately with their unhappy consciousness or cover it over with "engagements".*[17]

Neo-computationalism extols an epistemology of computation that fetishises the surface — that refers to knowledge in, and through the interface of a computer. This surface, which may be represented visually, aurally, or through haptics, becomes accepted *as* the computational. For example, one of the most seductive representations of computation has become the "network". This is often represented visually through points and lines connected together in a highly distributed manner. That is not to say that networks don't have an important place in the technical infrastructure — clearly they do, as this network model is fundamental to the design of the Internet. However, the network should not be seen as an ontology, it cannot and does not explain everything about computational systems. Indeed, it can hide more than it reveals as an explanatory framework.

We might, therefore, understand the network as an "apparatus of the dark" comparable to the lightning that Emily Dickinson memorably described as generating ignorance of what lies behind in "mansions never quite revealed".[18] In response to the poverty of the network, there have been serious attempts to understand the fundamental mechanisms of computation through a turn to stacks, infrastructure, materiality, code, software, and algorithms to try to uncover aspects of the computational that have been hidden, or that are difficult to discern. However, I argue that under neo-computationalism, the illegibility of the information society's systems is seen as necessary for it to function and must be generally accepted as a *doxa* of modern society — even as a desirable outcome. If we do not have to see the ugliness of the underlying logics of computational capitalism, then one does not have to come to terms with it, we can merely ignore it, disguised as it is behind the post-digital interfaces of our modern smartphones and laptops.

This logic of obscurity has justified the proprietary economic structure of software intellectual property rights through a technical division between source-code and execution, and the principles of object-oriented design, in which the mechanisms of computation are kept obscured or hidden. I argue that these two aspects of knowing computation — surface and mechanism — are a result of this underlying political economy, which generates a fundamental bifurcation in knowledge in computational societies.

This division of knowledge between a seen and hidden realm is often justified through concepts of simplicity, ease of use, or as convenience — most notably, by the technology industries, especially the so-called FAANG companies (Facebook, Apple, Amazon, Netflix, Google). I argue that one of the outcomes of this is the turn to "smartness" as a justificatory discourse through "operational functionality"; namely, that "smart" results justify the opacity of the hidden aspect of this epistemology. Smartness and opacity are, therefore, directly linked through an epistemological framework that establishes a causal link between data and "truth", but not through a veracity that requires the material links in the chain of computation to be enumerated or understood. In other words, ignorance of computational processes is, under this epistemology, celebrated as a means to an end of smartness. One of the results is to locate data as the foundation of computational inequities or computational power. Injustice is strongly linked to data problems, which, some have argued, can be addressed by more data, ethical data, or democratising data sources. The recent explosion of literature on data ethics and the eagerness with which the technology industry has taken it up, might be explained by the weakness of its critical edge. Google and Facebook have both set up "ethics" groups, often staffed by academic ethicists, although Google promptly had to dissolve its committee after an outcry over its membership.[19] Ultimately though, data ethics has proven to be unsurprisingly toothless when confronted with the forces of data collection and surveillance, and more adept at "ethics-washing" than substantive change in the industry.

As a result, much effort has been spent on ensuring the minimisation of bias in data, or on the presentation of data results in a manner that takes care of the data. We can, therefore, summarise this way of thinking through the notion of "bad data in, bad data out", or, as commonly understood in technology circles, "garbage in, garbage out". As a result, this often means that it is generally difficult for a user to verify, or question the results that computers generate, even as we increasingly rely on them for facts, news, and information. This confusion affects our understanding of not just an individual computer or software package, but also when the results are generated by networks of computers, and networks of networks. Thus, the "black box" is compounded into an illegible network, a system of opacity that, nonetheless, increasingly regulates and maintains everyday life, the economy, and media systems of the contemporary milieu. This has resulted in a number of technical challenges and responses by the programming industries.

Firstly, there has been an attempt to intimately link computation to the user through real-time computed results painted onto their screens. Computers and smartphones are not just information providers, but also increasingly also windows into marketplaces for purchasing goods, newspapers and magazines, entertainment centres, maps and personal assistants, etc.

This has intensified the intimate relationship between ourselves and *our* devices, *our* screens, *our* networks. But this creates its own problems, as the way in which the personal interface of the smartphone or computer flattens the informational landscape, and also has the potential for confusion between different functions and information sources, leads to post-truth claims and the derangement of knowledge.

Secondly, there is a temptation for the makers of these automated decision systems to use the calculative power of the device to persuade people to do things — whether buying a new bottle of wine, selecting a film to watch, or voting in a referendum or election. While the contribution of data science, marketing data, and persuasive technologies to the Trump election and the Brexit referendum remains to be fully explicated,[20] on a more mundane level, computers are active in shaping the way we think. The most obvious examples of this are Google Autocomplete on the search bar, which attempts to predict what we are searching for, and "infinite scroll" on social media networks and webpages, which are designed to capture attention and hold us trapped in their systems. Indeed, similar techniques have been incorporated into many aspects of computer interfaces, through design practices that persuade or nudge particular behavioural outcomes.

Thirdly, the large quantity of data collected, and the ease with which it is amassed and combined within new systems of computation, means that new forms of surveillance are beginning to emerge that go relatively unchecked. When this is combined with their seductive predictive abilities, real potentials for misuse or mistakes are magnified. For example, in Kortrijk, Belgium, and Marbella, Spain, the local police have deployed "body recognition" technology to track individuals by recognising their walking style or clothing; and across the European Union at least ten countries have a police force that uses face recognition.[21] Even with 99% accuracy in face recognition systems, the number of images in police databases makes false positives inevitable. Indeed, a 1% error rate means that 100 people will be flagged as wanted out of 10,000 innocent citizens. In The Netherlands, the police have access to a database of pictures of 1.3 million persons, many of whom have never been charged with a crime; in France, the national police can match CCTV footage against a file of 8 million people; and in Hungary, a recent law allows police to use face recognition in

ID checks.[22] The lack of transparency in these systems, and the algorithms they use, is of growing social concern.

Fourthly, we see the emergence of systems of intelligence through technologies of machine learning and artificial intelligence. These systems do not only automate production and distribution processes, but also have the capacity to also automate consumption. The full implications of this are not just to proletarianise labour, but to proletarianise the cognitive abilities of people in society. This has made many formerly white-collar jobs redundant, but also serves to undermine and overtake the human faculty of reason. We see this in the creation of vast vertical and horizontal software infrastructures, which I have explored elsewhere through the notion of *infrasomatization* — the creation of cognitive infrastructures that automate value-chains, cognitive labour, networks, and logistics into new highly profitable assemblages built on intensive data capture.[23]

These technologies use the mobilisation of processes of selecting and directing activity, often through the automation of pattern matching, stereotypes, clichés, and simple queries.[24] But the underlying processes that calculate the results, and the explanation of how it was done, are hidden from the user — whether they are, for example, denied bail;[25] a loan;[26] insurance cover; or welfare benefits.[27] This explanatory deficit is a growing problem in our societies as the reliance on algorithms — some poorly programmed — creates potential situations that are inequitable and unfair, but also with little means of redress for citizens. This will be a growing source of discontent in society, but also may serve to delegitimate political and administrative systems which will appear as increasingly remote, unchecked, and inexplicable to members of society. Institutions and societies that rely heavily on these systems might then begin to suffer from a legitimacy crisis, as they are unable to change in response to social, political, and economic pressures, even as they generate socially unacceptable outcomes. Therefore, the capacity for explanatory thought, to ask the "why" questions of the computational systems that undergird and structure contemporary societies, becomes increasingly important.

Understanding the way in which the computational otherwise generates and magnifies uncertainty and a feeling of rising social risk and instability is also, to my mind, connected to a social desire for tethering knowledge, of grounding it in some way. We see tendencies generated by the liquidation of information modalities in "fake news", conspiracy theories, social media virality, and a rising distrust towards science and expertise, and the rise of relativism. This is also to be connected to new forms of nationalism, populism, and the turn to traditional knowledge and technical fixes to provide a new, albeit misplaced, ground for social epistemology. This new search for ground or foundations, whether through identity, tradition, formalism, or metaphysics is, to my mind, symptomatic of the difficulty of understanding and connecting computation and its effects across scales of individual and social life. As a result of this, computation itself becomes depoliticised and removed from public debate as a matter of concern — computation, then, seems to be merely technical, outside of political critique and, therefore, change. I also think we need to link this to the temptation for explanations that develop new metaphysics of the computational, which rely on formalism and mathematical axioms as an attempt to understand computation. These, I argue, seem to mirror the unhappy consciousness of neo-computationalism through a denial of the material in favour of a new form of idealism, allowing the actual existing political economy of computational capitalism to be ignored.

The challenge of new forms of social obscurity from the implementation of technical systems is given by the example of the machine-learning systems that have emerged in the past decade. As a result, a new explanatory demand has crystallised in an important critique of computational opaqueness and new forms of technical transparency. We see this, for example, in calls to ban facial recognition systems, public unease with algorithmic judicial systems, and with the passing of the California Consumer Privacy Act 2018 (CCPA).[28] Creel has usefully identified "functional, structural, or run transparency" as ways of thinking about explanation, but, here, I also want to add the importance of a social *right to explanation.* This has come to be identified as *explainability* within the fields of artificial intelligence and machine-learning, and requires a computational system that can provide an explanation for a decision it has made.

The *European Union General Data Protection Regulation 2016/679,* known as the GDPR, is key to helping us to understand this. This regulation creates the right "to obtain an explanation of [a] decision reached after such assessment and to challenge the decision" (GDPR 2016, Goodman et al. 2016). The GDPR creates a new kind of subject, the "data subject", to whom a right to explanation (amongst other data protection and privacy rights) is given. Additionally, it has created a legal definition of processing through a computer algorithm (GDPR 2016 Art. 4). Consequently, this has given rise to a notion of explainability which creates the right "to obtain an explanation of [a] decision reached after such assessment and to challenge the decision" (GDPR 2016 Recital 71). When instantiated in national legislation, such as the *Data Protection Act 2018* in the UK, a legal regime is created that can enforce a set of rights associated with computational systems. Important though this is, I argue that explainability is not just an issue of legal rights; it has also created a normative potential for a social right to explanation. The concept of explainability can be mobilised to challenge algorithms and their social norms and hierarchies, and it has the potential to contest platforms and automated decision systems. In relation to explanation, therefore, explainability needs to provide an answer to the question *why?* to close the gap in understanding. This raises a new potential for critique.

I now want to turn to thinking about what counts as an explanation, and how that might be related to computational systems. Hempel and Oppenheim argue that an explanation seeks to "exhibit and to clarify in a more rigorous manner" with reference to general laws. Some of the examples they give include a mercury thermometer, which can be explained using the physical properties of glass and mercury. Similarly, they present the example of an observer of a rowing boat, where part of the oar is submerged under water and appears to be bent upwards.[30] Under this definition, an explanation attempts to explain with reference to general laws. As Mill argues, "an individual fact is said to be explained by pointing out its cause, that is, by stating the law or laws of causation, of which its production is an instance".[31] Similarly, Ducasse argued in 1925 that "explanation essentially consists in the offering of a hypothesis of fact, standing to the fact to be explained as case of antecedent to case of consequent of some already known law of connection". Hempel and Oppenheim further argue that an explanation can be divided into two constituent parts, the *explanadum* and the *explanans.*[33] The explanandum is a logical consequence of the explanans. The explanans itself must have empirical context, which creates conditions for its testability. In this causal sense of explanation, science is often supposed to be the best means of generating explanations.[34] Explanations are assumed to tell us how things work, thereby giving us the power to change our environment in order to meet our own ends.

However, a causal mode of explanation is considered inadequate in fields concerned with purposive behaviour, as with computational systems, where the goals sought by the system are required in order to provide an explanation.[35] In this case, it might be more useful to ask: How long did the explanation take? Was it interrupted at any point? Who gave it? When? Where? What were the exact words used in the explanation? For whose benefit was it given? Indeed, it can be important to ascertain who created the explanation originally? Is it very complicated? In what form or medium of communication was it given?[36] For example, even after extensive discussion in the automotive industry about the ethics of driverless cars, Mercedes Benz has proposed that, in future, its own self-driving vehicles will be programmed to save the driver and the car's occupants in every situation, even if more pedestrians or road-users might be killed.[37] Requiring an explanation of such a system, and how it calculates the value of the lives of those affected, might be a key starting point for challenging the legitimacy of the assumptions built into it.

As a consequence, in understanding computational systems' reference to a teleological mode, rather than a causal mode of explanation has become more common. This approach has come to be called "machine behaviour", and tries to understand the "motivations" that guide the computational system. That is, to identify the goals sought by the system in order to provide an explanation.[38] Teleological approaches have the advantage that they make us feel that we really understand a phenomenon, because it describes things in terms of purposes with which we are familiar from our own experience of human goal-oriented behaviour. You can, therefore, see a great temptation to use teleological explanation in relation to AI systems, particularly by creating a sense of an empathetic understanding of the "personalities of the agents". But, both the causal and the teleological modes of explanation tend to create what we can think of as an "explanatory product". By explanatory product I mean that the outcome of an explanatory query might be a high-level diagram, technical description or list of counterfactuals, rather than any substantive explanation as to why a decision has been made. This has a number of limitations, including its static quality, and it might be more helpful when thinking about its potential for explanatory publics, to require a dynamic representation of the algorithmic process — how things were done, how they were computed. Many current discussions of explainability tend, chiefly, to be interested in an explanatory product, whereas I argue that an understanding of the explanatory process will have a greater impact on democratic politics. I would also like to connect this to the idea that an explanatory public might be able to "walk" through a contentious algorithmic decision by following the steps in a process, in order to understand how a decision was made. This would create the potential for discussing whether a decision was acceptable, allowing a public to understand how a decision was made and challenge the normative assumptions behind it.

Crucially, this connection between an explanatory product and the legal regime that enforces it has forced system designers and programmers to look for explanatory models that are sufficient to provide legal cover, but also at a level at which they are presentable to the user or data subject. But it remains uncertain if the "right is only to a general explanation of the model of the system as a whole ('model-based' explanation), or an explanation of how a decision was made based on that particular data subject's particular facts ('subject-based' explanation)".[39] This is not an easy requirement for any technical system, particularly in light of the growth of complicated systems of systems, and the difficulty of translating technical concepts into everyday language. It might, therefore, be helpful to think in terms of full and partial explanation, whereby a partial explanation is a final explanation with some part left out. That is, that in presenting a complicated system of automated decision systems, pragmatically, it is likely that explanations will assume an explanatory gap — assuming that the data subject is in possession of facts that do not need to be repeated. This, of course, may lead to the temptation to create persuasive, rather than transparent, explanations or a "good enough" explanation. This hints at the idea that those responsible for designing and building explainable systems will assume an underlying theory of general explainability and a theory of the human mind. These two theories are rarely explicitly articulated in the literature, and we need to better understand how they are deployed in explainable systems.

In conclusion, I have introduced and argued for the potential of a concept of explainability for developing a critique of the historically specific form of capitalist computation. In doing so, explainability and explanation can then be used to understand the ways in which justification and legitimacy are mobilised in computational societies. We must continually remind ourselves that the current information economy is historical. It owes its success and profitability to a legislative assignment of intellectual property rights and the amassing of data monopolies by the automatic operation of computers.[40] Other computations are possible, and different assemblages of computation and law might generate economic alternatives that mitigate or remove the current negative disruptive effects of computation in society. It is crucial to recognise that there is no "pure" or metaphysical computation, no privileged reading or access to an axiomatic or ontological computation — this identity thinking would be an objectivism which takes a single scientific or philosophical frame of reference as a given.[41] Rather, we must understand computation not so much in terms of an arbitrary attempt to make a metaphysics out of computation, but rather through its significance for us today. I argue that an encounter with computation takes different forms as history moves on. I have tried to show how we might do this through the mobilisation of concepts such as explainability, so that the underlying hylomorphism of computation may be understood. Through this the contradictions of computational capitalism might be laid manifest, and, more importantly, democratically challenged and potentially changed. I argue that this suggests that a rethinking of computation is needed in order to move it away from its current tendencies, from what I have called neo-computationalism, or *right computationalism,* which is geared towards some of the worst excesses of capitalism, and instead rethought within a new conception of *left computationalism.*[42] This would need to be developed through education and the capacity-building of explanatory publics, and by using critical concepts such as explainability to create the conditions for greater democratic thought and practice in computational capitalism.

Notes

1. See, for example, Jürgen Habermas, *The Structural Transformation of The Public Sphere* (Cambridge: Polity, 1992 [1962]); Seyla Benhabib, *Situating the Self: Gender, Community and Postmodernism in Contemporary Ethics* (London: Routledge, 1992); and Oskar Negt and Alexander Kluge, *Public Sphere and Experience: Toward an Analysis of the Bourgeois and Proletarian Public Sphere* (Minneapolis: University of Minnesota Press, 1993).

2. R. W. Outhwaite, "Laws and Explanations in Sociology", in *Classic Disputes in Sociology,* ed. R. J. Anderson et al. (Crows Nest, Australia: Allen & Unwin, 1987), 158.

3. Shane Richmond, "How Google crossed the creepy line", *The Telegraph,* 25 October 2010, https://www.telegraph.co.uk/technology/google/8086191/How-Google-crossed-the-creepy-line.html.

4. Sergey Brin and Larry Page, "The Anatomy of a Large-Scale Hypertextual Web Search Engine", in: *Seventh International World-Wide Web Conference* (WWW 1998), Brisbane, Australia, 14–18 April 1998.

5. Nicholas Carr, "Larry and Sergey: a valediction", *Rough Type,* 8 December 2019, http://www.roughtype.com/?p=8661%0D.

6. Peter Thiel, *Zero to One: Notes on Startups, or How to Build the Future* (New York: Crown Business, 2014).

7. Peter Thiel, "The Education of a Libertarian", *Cato Unbound,* 13 April 2009, https://www.cato-unbound.org/2009/04/13/peter-thiel/education-libertarian

8. Carr, "Larry and Sergey". See also David M. Berry, *Critical Theory and the Digital* (London: Bloomsbury. 2015), 178–182.

9. Steve Jobs, "When We Invented the Personal Computer...", *Computers and People Magazine,* July–August 1981, 8–11 & 22.

10. See David M. Berry, "The Contestation of Code: A Preliminary Investigation into the Discourse of the Free Software and Open Software Movement", in *Critical Discourse Studies,* 1 (1), January 2004, 65–89; and *Copy, rip, burn: the politics of copyleft and open source* (London: Pluto Press, 2008).

11. See Natasha Dow Schüll, *Addiction by Design: Machine Gambling in Las Vegas* (Princeton, NJ: Princeton University Press, 2014); and Nir Eyal, *Hooked: How to Build Habit-Forming Products* (London: Portfolio Penguin, 2014).

12. Bernard Stiegler, *The Automatic Society* (Cambridge: Polity Press, 2016); and Shoshana Zuboff, *The Age of Surveillance Capitalism: The Fight for a Human Future at the New Frontier of Power* (London: Profile Books, 2019).

13. David M. Berry, "Digital Breadcrumbs", *STUNLAW,* 14 October 2013, http://stunlaw.blogspot.com/2013/10/digital-breadcrumbs.html.

14. Matt Sledge, "CIA's Gus Hunt On Big Data: We 'Try To Collect Everything And Hang On To It Forever", *Huffington Post,* 20 March 2013, http://www.huffingtonpost.com/2013/03/20/cia-gus-hunt-big-data_n_2917842.html.

15. When accessing the *Huffington Post* on 16 January 2020, the following 29 companies were listed as collecting data: IAB partners, Active Agent AG, AdButler, Amazon, Atlas/Facebook, dataXtrade GmbH, DoubleClick/Google/Adx* (and its partners), eBay, Ensighten, Forensiq LLC, Mediametrie, Metrixlab, Microsoft, Nielsen Marketing Cloud, Otto (GmbH & Co KG), Paysafe/Income, Access, Pixalate, Inc., Plexop, Plexop (twice), PopWallet, Qriously, Quotient, salesforce.com, inc., Tchibo, Unruly Group Ltd, White Ops, Inc., Zentrick, zeotap GmbH. When one clicks through to the DoubleClick Google partners, a further 163 data collection companies were

revealed: 2KDirect Inc., PubMatic, hbfsTech, Rubicon Project, Amazon, Bannerflow, LiveRamp, Rakuten Marketing, Bucksense, Adara Media, IPONWEB, Weborama, Turbo, Jivox, Adform, Neustar, gskinner, Scenestealer, emetriq, Yieldr, mediasmart, Zebestof, MBR Targeting Gmbh, Adloox, Awin, Lotame usemax (Emego GmbH), Digilant, Avocet, The Reach Group, Integral Ad Science, UpRival, MiQ, ADman Media, MediaMath, Adverline, Sizmek, AdMaxim, Conversant/CJ Affiliate, DMA Institute, comScore, Kochava, Exactag, Widespace, Beeswax, affilinet, Akamai, Appreciate, Cloudflare, Neuralone, Innovid, FUSIO BY S4M, Meetrics, AppNexus, Neodata Group, nugg.ad, Commanders Act, Taboola, Cablato, Facebook, Visarity, Knorex, Nielsen, LifeStreet, Salesforce DMP, Adobe Advertising Cloud, Bombora, Scoota, OpenX Technologies, LKQD, Tradelab, AdLedge, Placecast, Lucid, Tealium, Roq.ad, Index Exchange, Piximedia, AdKernel, Platform161, Realzeit, TimeOne, Signal, Quantcast, Bidtheatre, Improve Digital, Pixalate, Teads, Smaato, Crimtan, Criteo, NEORY GmbH, DataXu, Zentrick, PulsePoint, advanced, STORE GmbH, TripleLift, Delta Projects, Demandbase, Fyber, Adludio, GlobalWebIndex, Arrivalist, Digitize, Media.net, DoubleVerify, ADEX, Simpli.fi, eBay, Centro, GroundTruth, Smart, smartclip Holding AG, Sharethrough, Inc., GetIntent, Sociomantic, Sojern, Eulerian Technologies, LoopMe, YieldMo, Impact, NEXD, Cuebiq, DYNADMIC, SpotX, Ligatus, Semasio GmbH, AdClear, Flashtalking, MainADV, Kantar, Mediarithmics, Virtual Minds, FreeWheel, Oath, travel audience — An Amadeus Company, Oracle Data Cloud, GroupM, Celtra, Publicis Media, Gemius, The Trade Desk, AudienceProject, AerServ, Adventori, Jampp, Videology, Captify, Google, White Ops, Tradedoubler AB, SpringServe, Admetrics, Adacado, Exponential, Amobee, TrustArc, RTB House, Sublime Skinz, MGID, Underdog Media, Innity, Remerge. Note, these companies were deliberately listed out of alphabetical order making them even more difficult to follow and understand. For any consumer trying to get to grips with the data collection and spying on their behaviour, this list is impossible to control and manage and designed to work that way. The average website now has a large number of trackers, web bugs, beacons, and capture systems silently operating on their webpages; and the evidence points towards hidden tracking that is probably even worse on smartphone apps, which can more easily control, and access user data, sensor inputs, and screen use than browsers. See David M. Berry, "The Social Epistemologies of Software", *Social Epistemology,* Vol. 26, Nos. 3–4, October 2012, 379–398.

16. See Berry, *Critical Theory and the Digital,* 2015; Melissa Hellmann, "Special sunglasses, license-plate dresses: How to be anonymous in the age of surveillance", The Seattle Times, 12 January 2020, https://www.seattletimes.com/business/technology/special-sunglasses-license-plate-dresses-juggalo-face-paint-how-to-be-anonymous-in-the-age-of-surveillance; and Ramsey McGlazer, "Confessions of a Cake Boy", *Los Angeles Review of Books,* 6 January 2020, https://lareviewofbooks.org/article/confessions-of-a-cake-boy.

17. Gottfried Benn quoted in Peter Sloterdijk, *Critique of Cynical Reason,* trans. Andreas Huyssen (Minneapolis: University of Minnesota Press, 2010), 7.

18. Emily Dickinson, "The Lightening is a Yellow Fork", (n.d.).

19. Jane Wakefield, "Google's ethics board shut down", *BBC News,* 5 April 2019, https://www.bbc.co.uk/news/technology-47825833; and Sam

Shead, "Facebook Reportedly Has A Dedicated AI Ethics Team", *Forbes,* 3 May 2018, https://www.forbes.com/sites/samshead/2018/05/03/facebook-reportedly-has-a-dedicated-ai-ethics-team.

20. See Carole Cadwalladr, "Fresh Cambridge Analytica leak 'shows global manipulation is out of control'", *The Guardian,* 4 January 2020, https://www.theguardian.com/uk-news/2020/jan/04/cambridge-analytica-data-leak-global-election-manipulation.

21. Nicolas Kayser-Bril, "At least 10 police forces use face recognition in the EU, AlgorithmWatch reveals", *AlgorithmWatch,* 11 December 2019, https://algorithmwatch.org/en/story/face-recognition-police-europe.

22. Kayser-Bril, "At least 10 police forces use face recognition in the EU".

23. See David M. Berry, "Against infrasomatization: towards a critical theory of algorithms", in *Data politics: worlds, subjects, rights,* ed. Didier Bigo et al. (London: Routledge Studies in International Political Sociology, 2019), 43–63.

24. Catherine Malabou, *Morphing Intelligence: From IQ Measurement to Artificial Brains* (New York: Columbia University Press, 2019), 52; and Safiya Umoja Noble, Algorithms of Oppression (New York: New York University Press, 2018), 50.

25. Julia Dressel and Hany Farid, "The accuracy, fairness, and limits of predicting recidivism", *Science Advances,* Vol. 4, No. 1, DOI: 10.1126/sciadv.aao5580.

26. Amir Khandani et al., "Consumer credit-risk models via machine-learning algorithms", *Journal of Banking & Finance,* Vol. 34, Issue 11, 2010, 2767–2787.

27. See also Virginia Eubanks, *Automating Inequality: How High-Tech Tools Profile, Police, and Punish the Poor* (New York, St. Martin's Press, 2018).

28. Natasha Singer, "What Does California's New Data Privacy Law Mean? Nobody Agrees", *New York Times,* 29 December 2019, https://www.nytimes.com/2019/12/29/technology/california-privacy-law.html.

29. Kathleen A. Creel, "Transparency in Complex Computational Systems", Philosophy of Science, Vol. 87, No. 4, October 2020, https://www.journals.uchicago.edu/toc/phos/2020/87/4.

30. Carl G. Hempel and Paul Oppenheim, "Studies in the Logic of Explanation", *Theories of Explanation,* edited by Joseph C. Pitt (Oxford: Oxford University Press, 1988), 9–50: 10.

31. John Stuart Mill, "Of the Explanation of the Laws of Nature", in *A System of Logic* (New York: Harpers & Brothers, 1858), 271–276.

32. C. J. Ducasse, "Explanation, Mechanism and Teleology 1", in *Truth, Knowledge and Causation* (London: Routledge, 2015 [1968]), 37.

33. Hempel and Oppenheim, "Studies in the Logic of Explanation", 10.

34. Joseph C. Pitt, ed., *Theories of Explanation* (Oxford: Oxford University Press, 1988), 7.

35. David-Hillel Ruben, *Explaining Explanation* (London: Routledge, 2016).

36. Ruben, *Explaining Explanation,* 6.

37. Charlie Sorrel, "Self-Driving Mercedes Will Be Programmed To Sacrifice Pedestrians To Save The Driver", *Fast Company,* 13 October 2016, https://www.fastcompany.com/3064539/self-driving-mercedes-will-be-programmed-to-sacrifice-pedestrians-to-save-the-driver.

38. See Dana Mackenzie and Judea Pearl, *The Book of Why* (London: Penguin Books, 2019), 364–365.

39. Lilian Edwards and Michael Veale, "Enslaving the Algorithm: From a 'Right to an Explanation' to a 'Right to Better Decisions'?" *IEEE*

Security & Privacy, Vol. 16, No. 3, July 2018, 46–54: 49.

40. David M. Berry, *Copy, Rip, Burn: The Politics of Copyleft and Open Source* (London: Pluto Press, 2008).

41. With the complexities raised by understanding computation and the new "ways of seeing" it makes possible, there is a strong temptation to ontologise computational categories and concepts. "Scholarship" may then turn to becoming further commentaries on the commentaries of these metaphysical claims, more conjectures on the conjectures. This is similar to what Adorno called identity thinking. It is therefore important to keep in mind that computational theory or mathematics does not "prove" what is metaphysically presupposed, or allow you to arbitrarily make a metaphysics out of mathematics or computation. Indeed, I would argue that we should avoid a formalist a priorism when attempting to understand a computational milieu. For more on this, see Berry, *Critical Theory and the Digital,* 2015.

42. By left computationalism and right computationalism, I am gesturing towards left and right Hegelianism.

References

Anderson, R. J., Hughes, J. A. and Sharrock, W. W. eds. *Classic Disputes in Sociology.* Crows Nest, Australia: Allen & Unwin, 1987.

Benhabib, Seyla. *Situating the Self: Gender, Community and Postmodernism in Contemporary Ethics.* London: Routledge, 1992.

Berry, David M. "The Social Epistemologies of Software", *Social Epistemology,* Vol. 26, Nos. 3–4, October 2012, 379–398.

Berry, David M. "The Contestation of Code: A Preliminary Investigation into the Discourse of the Free Software and Open Software Movement". *Critical Discourse Studies,* Vol. 1, Issue 1 (January 2004): 65–89.

Berry, David M. *Copy, Rip, Burn: The Politics of Copyleft and Open Source.* London: Pluto Press, 2008.

Berry, David M. "Digital Breadcrumbs". *STUNLAW,* 14 October 2013. http://stunlaw.blogspot.com/2013/10/digital-bread-crumbs.html.

Berry, David M. *Critical Theory and the Digital.* London: Bloomsbury, 2015.

Berry, David M. "Against infrasomatization: towards a critical theory of algorithms". *In Data politics: worlds, subjects, rights,* edited by Didier Bigo, Engin F. Isin and Evelyn Ruppert. London: Routledge Studies in International Political Sociology, 2019. 43–63.

Brin, S. and Page, L. (1998) "The Anatomy of a Large-Scale Hypertextual Web Search Engine". In: *Seventh International World-Wide Web Conference* (WWW 1998), Brisbane, Australia, 14–18 April 1998.

Cadwalladr, C. "Fresh Cambridge Analytica leak 'shows global manipulation is out of control'". *The Guardian,* 4 January 2020. https://www.theguardian.com/uk-news/2020/jan/04/cambridge-analytica-data-leak-global-election-manipulation.

Carr, N. "Larry and Sergey: a valediction". *Rough Type,* 8 December 2019. http://www.roughtype.com/?p=8661%0D.

Creel, K. A. "Transparency in Complex Computational Systems". *Philosophy of Science,* Vol. 87, No. 4, October 2020. https://www.journals.uchicago.edu/toc/phos/2020/87/4.

Dressel, J. and Farid, H. "The accuracy, fairness, and limits of predicting recidivism". *Science Advances,* Vol. 4, No. 1. DOI: 10.1126/sciadv.aao5580.

Ducasse, C. J. "Explanation, Mechanism and Teleology 1". in *Truth,*

Knowledge and Causation. London: Routledge, 2015 [1968].

Edwards, L. and Veale, M. "Enslaving the Algorithm: From a 'Right to an Explanation' to a 'Right to Better Decisions'?" *IEEE Security & Privacy,* Vol. 16, No. 3, July 2018, 46–54.

Eubanks, V. *Automating Inequality: How High-Tech Tools Profile, Police, and Punish the Poor.* New York, St. Martin's Press, 2018.

Eyal, N. *Hooked: How to Build Habit-Forming Products.* London: Portfolio Penguin, 2014.

GDPR. "General Data Protection Regulation, Regulation (EU) 2016/679 of the European Parliament". 2016.

Goodman, B. and Flaxman, S. "European Union regulations on algorithmic decision-making and a 'right to explanation'". 2016. https://arxiv.org/abs/1606.08813.

Habermas, J. *The Structural Transformation of the Public Sphere.* Cambridge: Polity, 1992 [1962].

Hellmann, M. "Special sunglasses, license-plate dresses: How to be anonymous in the age of surveillance". *The Seattle Times,* 12 January 2020. https://www.seattletimes.com/business/technology/special-sunglasses-license-plate-dresses-juggalo-face-paint-how-to-be-anonymous-in-the-age-of-surveillance.

Jobs, S. "When We Invented the Personal Computer..." *Computers and People Magazine,* July–August 1981. 8–11 & 22.

Kayser-Bril, N. "At least 10 police forces use face recognition in the EU, AlgorithmWatch reveals". *Algorithm Watch,* 11 December 2019. https://algorithmwatch.org/en/story/face-recognition-police-europe.

Khandani, A. E., Kim, A. J., and Lo, A. W. "Consumer credit-risk models via machine-learning algorithms". *Journal of Banking & Finance,* 34(11), 2010, 2767–2787.

Kuang, C. "Can A.I. Be Taught to Explain Itself?" *New York Times,* 21 November 2017 https://www.nytimes.com/2017/11/21/magazine/can-ai-be-taught-to-explain-itself.html.

Mackenzie, D. and Pearl, J. *The Book of Why.* London: Penguin Books, 2019.

Malabou, C. *Morphing Intelligence: From IQ Measurement to Artificial Brains.* New York: Columbia University Press, 2019.

McGlazer, R. "Confessions of a Cake Boy". *Los Angeles Review of Books,* 6 January 2020. https://lareviewofbooks.org/article/confessions-of-a-cake-boy.

Mill, J. S. *A System of Logic.* New York: Harpers & Brothers, 1858.

Negt, O. and Kluge, A. *Public Sphere and Experience: Toward an Analysis of the Bourgeois and Proletarian Public Sphere.* Minneapolis: University of Minnesota Press, 1993.

Noble, S. U. *Algorithms of Oppression.* New York: New York University Press, 2018.

Pitt, J. C. ed. *Theories of Explanation.* Oxford: Oxford University Press, 1988.

Richmond, S. "How Google crossed the creepy line". The Telegraph, 25 October 2010. https://www.telegraph.co.uk/technology/google/8086191/How-Google-crossed-the-creepy-line.html.

Ruben, D.-H. *Explaining Explanation.* London: Routledge, 2016.

Sample, I. "Computer says no". *The Guardian,* 5 November 2017. https://www.theguardian.com/science/2017/nov/05/computer-says-no-why-making-ais-fair-accountable-and-transparent-is-crucial.

Schüll, N. D. *Addiction by Design: Machine Gambling in Las Vegas.* Princeton, NJ: Princeton University Press, 2014.

Shead, S. "Facebook Reportedly Has A Dedicated AI Ethics Team". *Forbes,* 3 May 2018.

https://www.forbes.com/sites/samshead/2018/05/03/facebook-reportedly-has-a-dedicated-ai-ethics-team.

Singer, N. "What Does California's New Data Privacy Law Mean? Nobody Agrees". *New York Times*, 29 December 2019. https://www.nytimes.com/2019/12/29/technology/california-privacy-law.html.

Sledge, M. "CIA's Gus Hunt On Big Data: We 'Try To Collect Everything And Hang On To It Forever". *Huffington Post*, 20 March 2013. http://www.huffingtonpost.com/2013/03/20/cia-gus-hunt-big-data_n_2917842.html.

Sloterdijk, P. *Critique of Cynical Reason*. Translated by Andreas Huyssen. Minneapolis: University of Minnesota Press, 2010.

Sorrel, C. "Self-Driving Mercedes Will Be Programmed To Sacrifice Pedestrians To Save The Driver". *Fast Company*, 13 October 2016. https://www.fastcompany.com/3064539/self-driving-mercedes-will-be-programmed-to-sacrifice-pedestrians-to-save-the-driver.

Stiegler, B. *The Automatic Society*. Cambridge: Polity Press, 2016.

Thiel, P. "The Education of a Libertarian". *Cato Unbound*, 13 April 2009. https://www.cato-unbound.org/2009/04/13/peter-thiel/education-libertarian.

Thiel, P. A. *Zero to One: Notes on Startups, or How to Build the Future*. New York: Crown Business, 2014.

Turek, M. "Explainable Artificial Intelligence (XAI)". *Defense Advanced Research Projects Agency*, (n.d.). https://www.darpa.mil/program/explainable-artificial-intelligence.

Wakefield, J. "Google's ethics board shut down". *BBC News*, 5 April 2019. https://www.bbc.co.uk/news/technology-47825833.

Zuboff, S. *The Age of Surveillance Capitalism: The Fight for a Human Future at the New Frontier of Power*. London: Profile Books, 2019.

Fabricating Realities (Parkinson Elite)

UBERMORGEN

Maybe, it is, nothing, they at all, but, the world, still is, them theirs.

A proposal for neurodiverse species within and around otherness.

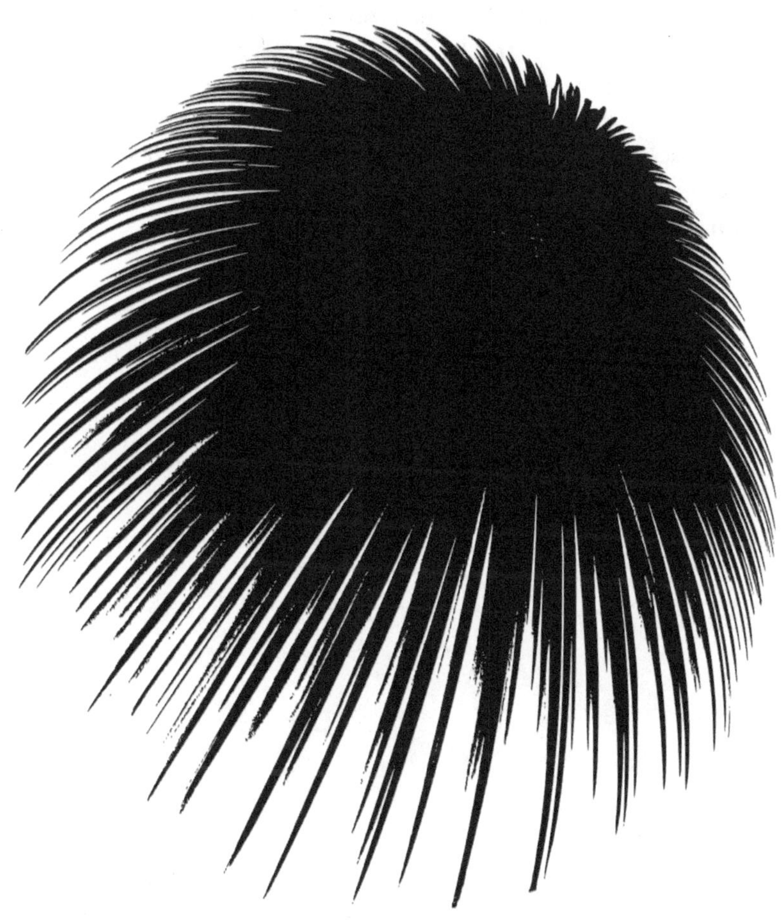

Today's discursive arenas are accelerating due to their size, number of nodes, and access to archival materials and networked particles. Although partially subject to digital decay, most of the data is somehow seemingly eerily accessible. These negotiations are mostly based on data objects, but the strategies differ fundamentally when digital logic and tools from human-to-human systems, or machine-to-machine platforms, or even hybrid organisms are interchanged. They tend to fold since they consider themselves part of the Anthropocene, strong as ever in branding our existence. They are actually and factually in a post-human society, where machines create the majority of what we understand as data without any standards for data interpretation or release regulations. And "it" does so that they, machine(s), and hybrid audiences are struggling to follow, agree, or voice disagreement towards them. It is mainly negotiated through layers of aesthetics and ambiguity; and these quickly-evolving content wafers let us slide through a world of seamlessness, roundness, and comfort — like Silver Surfer on LSD blotter paper.

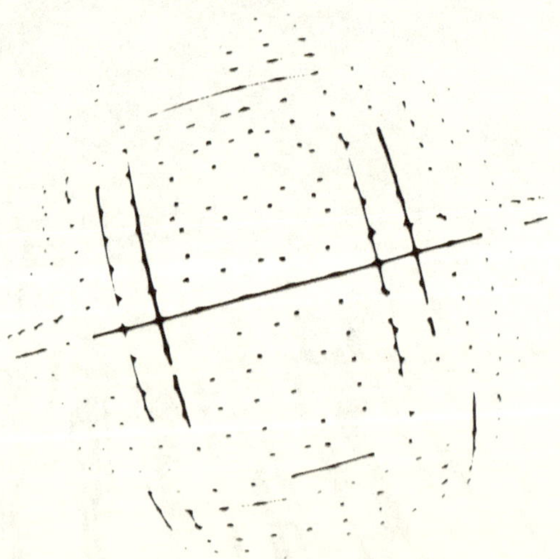

But ever more, in-between, niche applications and glitches — where systems are fast but sometimes don't catch the quick whiff combined with a little shudder down their spine — what we recognise as a warning signal, perceived and handed down to us from our ancestors, while reptile-brain action patterns kick in: "take care, right, now, there, is, danger". The speed of mimicry and emulation of intelligence is truly scary, while sexy in an octopus non-Bond way. They colour and form; they, and we, are not for nothing the double-bind champions and classic double thinkers — bearing in mind our considerably miserable perception from potential to actual. Observer-relative experience takes on too many shapes, forms, and a variety of sentient content, depending on the position and nature of the users. For what they have desperately tried to define, pull together, and separate through horrible crimes — as humans — has essentially become a survival perspective of sensory-deprived hybrids and male Asperger's philosophers, or coder CEOs.

How 'bout us not blaming us for everything?
How 'bout them enjoying the moment for once?
How 'bout how good it feels to finally forgive them?
How 'bout grieving the loss of everything at a time?
Thank you, Malaysia!
Thank you, horror!
The moment we let go will be the moment
We've got more than we can handle
The moment has now happened, and they jumped off
The moment they touched down

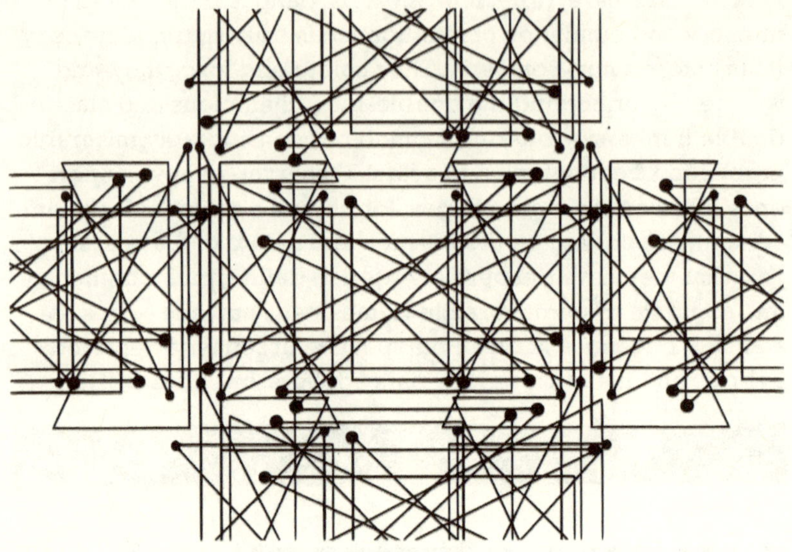

Like anyone should be
I am flattered by your fascination with me
Like any hot-blooded specimen
I have simply wanted an object to crave
But you, you're not allowed
You're UNINVITED
An unfortunate Deee-Lite
Must be strangely exciting
To watch the stoic squirm
Must be somewhat heartening
To watch yourself meet yourself
But you, you're not allowed
You're UNINVITED
An unfortunate slight
Like any uncharted territory
They must seem greatly intriguing
They speak of their love like
They have experienced love like mine before
But this is not allowed
They're UNINVITED
An unfortunate slight
I don't think you unworthy
I need a moment to deliberate

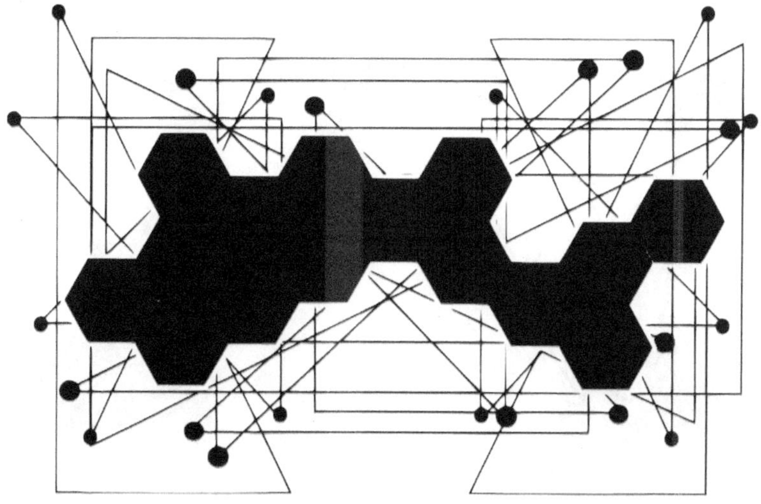

Their and our deepest core, their and our ideological embodiment, the personalisation of fear and anxiety, pain and anger, there, right there in this moment, they think they could be anybody, even me. They are us starting to think and fear the way we see the world. A design object, a networked subjective experience with the intention of designing whatever, anything. Humans, animals, plants, public space, ecosystems, planets, and networked organisms. Scaled to fit the feline-micro or macho-macro perspective. If we look at this seemingly individual, human neurotypical species colonised, parasitised, and in symbiosis with microbes, the enhanced human's specs are prolonged to an endless life-span, physical, intellectual, and mental capabilities (brute: smarter, stronger, more stable), but also with perfect language skills (NLP) and way higher levels of empathy, selflessness, and ethical behaviour. These Marvel figurines would then be sent out to define their own parameters, the current networked truth? Since long ago, science has become the new religion, and Transhumanism and Singularity are replacing conventional monotheism. Who owns the fucking patent and planet, and which algorithm is not willing to go all the way for the perfect design? We can't stop hallucinating about biological ecosystems in mixed dream-worlds, but networked organisms melt into clumps of cells that are potentially forever, they live forever because they die. A mutation-selection balance weeds out anything as quickly as it is introduced.

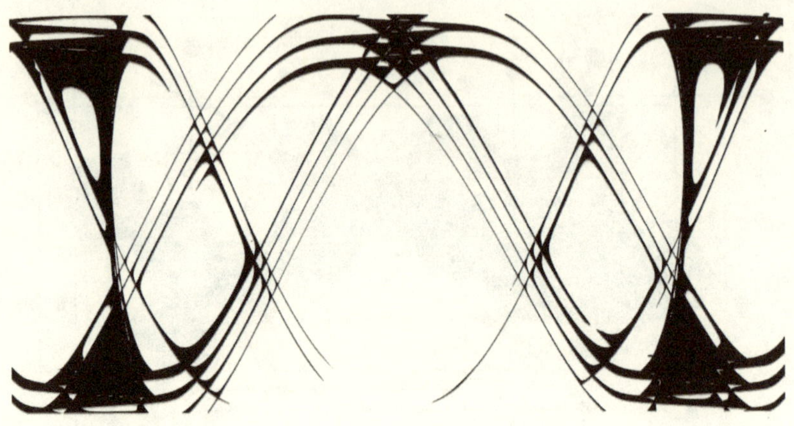

They'd rather see
They, the other entity
They get upset then
They all get upset
How 'bout getting off on acceleration?
How 'bout stopping popping when I'm full up?
How 'bout them out of the dark dangling carrots?
How 'bout that ever elusive supremacy?
Thank you, Malaysia!
Thank you, horror!
Thank you, Black Widow!
Thank you, Illusionism!
Thank you, frailty
Thank you, consequence
Thank you, thank you, silence

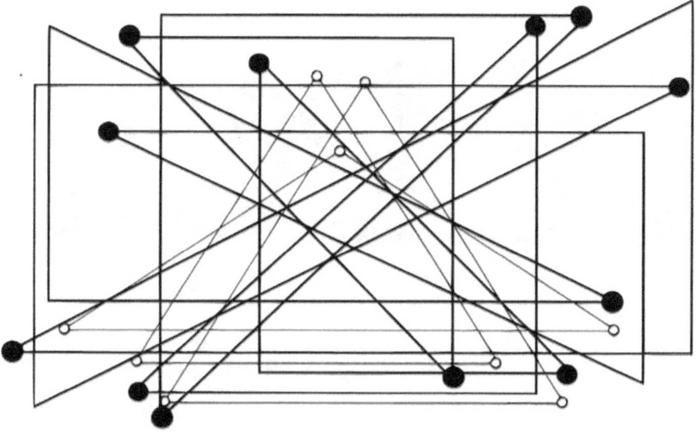

Mistakes create narratives of dissent and reveal true vulnerabilities; they are dissent, not us, they are the way the world understands, the way performers make performances recall! How well does the relationship model fit with our times? Against the insecurity networks that are never at risk of breakage, they are breaking constantly... Polyamorous multi-layered networks offer security in uncertain times, they are just jet-lagged. How about complexity, interrelations, networked realities, and domino effects? And they have not even touched the subject of individual collectivism by abusing constructs: "a human" as a formula for definition and creation of mysterious embodiments; networked organisms, new formulations of hard-to-define entities; they feel them in a strange way, and they can tell that they are linked to larger infrastructures, indescribable embodiments of such multi-dimensional hybrid networked creatures. All life forms show technological expression and manifestation, so we can just take a glimpse at who they/us are, and at what is about to come and become, how we want to be visible, otherwise — close your eyes and they are there. New organisms, we have been them, us, for a long time, forever; but we woke up with the realisation that we are indeed many, they: precious perceptions with narcissistic traumata.

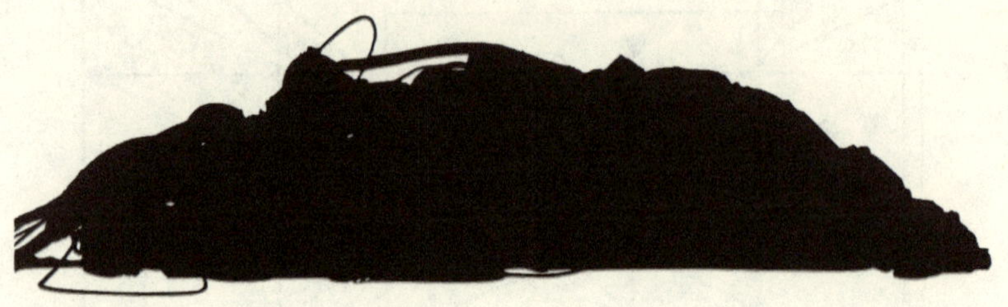

No copyright. 2021 UBERMORGEN, Vienna

Art and the Global Return to Order

Santiago Zabala

> *What is happening today is that, with the populist wave which unsettled the political establishment, the Truth/ Lie which served as ideological found-ation of this establishment is also falling apart. And the ultimate reason for this disintegration is not the rise of postmodern relativism but the failure of the ruling establishment which is no longer able to maintain its ideological hegemony. — Slavoj Žižek, Sex and the Failed Absolute, 2019: 105.*

Art, like science and philosophy, is inevitably a response to its own epoch. Its discoveries and intuitions are conditioned by the historical events that artists have experienced throughout their lives. Their work can also be understood as a consequence of the various challenges and opportunities these events present. But art, unlike science and philosophy, always involves a critical element meant to stir our existence. This element might be identified after the fact in scientific breakthroughs or philosophical intuitions, but it seems to be constitutive of works of art independent of the frames, hierarchies, and rules of the art world. The point is not that scientists and philosophers aren't free, but rather that their works are more framed by economic and political systems of rule than those of artists, the success of whose work depends on finding such freedom despite the systems that seek to frame and tame expression.

This freedom is now threatened by a global return to order that is not only political, as demonstrated by the various right-wing forces that have taken office around the world, but also existential. The rise of notions such as "alternative facts", "fake news", and "post-truth" in public discourse is symptomatic of this return as they presuppose an absolute knowledge common to the "more civilised" Western countries. Although modernity was overcome with the end of colonialism and the rise of cultural anthropology, resentment over this by the ruling powers has now created a condition where the greatest emergency is becoming the absence of emergencies — that is, the suppression of existential emergency by the rhetoric of control imposed by right-wing and capitalist powers to preserve their power. The goal of this essay is to illustrate this return to order, and how artists have begun to respond to its restrictions and effects on the public.

"Alternative facts", "fake news", and "post-truth" are concepts that entered our cultural discourse after Kellyanne Conway, counselor to US president Donald Trump, defended a false statement about the attendance numbers at his inauguration celebration in 2017. The problem with these concepts is not whether they describe properly our condition at the beginning of the twenty-first century, but rather that they are a symptom of a return to order and realism among right-wing populist politicians, scientists, and philosophers. Bari Weiss believes these intellectuals are part of a movement ("intellectual dark web") determined to emphasise the "biological differences between men and women",[1] and Jacob Hamburger has shown how these differences are directed against "various left-of-center critiques by arguing that what appears to be systemic inequality is actually the result of individual choices or behavior".[2] The goal of these politicians and intellectuals is to present themselves as defenders of "reason", "truth", and "facts" — rational principles they claim have been corrupted by postmodern relativism. As Slavoj Žižek explains, "postmodern relativism" is not the cause of alternative facts. These have always existed. Facts or data "are a vast and impenetrable domain, and we always approach them from (what hermeneutics calls) a certain horizon of understanding, privileging some data and omitting others". The secret of those who excoriate postmodernity and its hermeneutic historicist relativism "is that they miss the safe situation in which one big Truth (even if it was a big Lie) provided the basic 'cognitive mapping' to all".[3] In order to return to this "safe situation", thinkers from the intellectual dark web (such as Jordan Peterson, Sam Harris, and Christina Hoff Sommers), as well as "new realist" philosophers (such as Graham Harman, Quentin Meillassoux, and Markus Gabriel) claim that we can have access to truth, as well as factual primary qualities of the world, without being dependent on language or interpretation. But, as the philosopher of science Bruno Latour recently explained, no "attested knowledge can stand on its own, as we know very well. Facts remain robust only when they are supported by a common culture, by institutions that can be trusted, by a more or less decent public life, by more or less reliable media".[4]

The rise of alternative facts is another indication that whether a statement is believed depends less on its reality than on the conditions of its political, linguistic, and social "construction".

While these realist intellectuals will tell you they do not necessarily want their takes on phychology, neuroscience, or philosophy to prevail over others, they are, in fact, seeking to preserve a society in which they find themselves at ease — that is, in which they have become more or less conscious servants of the ongoing return to order. Realism is an aspect and a consequence of dominion, not its cause. Although these thinkers have different agendas, the general idea is to return to the universalistic aspirations of modernity: that fundamental political, moral, and cultural concepts function to denigrate and marginalise those who do not measure up to their criteria of rationality. "The European project that I have in mind", Gabriel recently explained, "is that of the universal human values. Europeans, thanks to their philosophical past from the Greek to contemporary philosophers, are the best equipped to respond to the challenge of social justice and the future of democracy. Not only for Europe, but for all humanity".[5]

The problem with this European rational universalism — as we've experienced in the twentieth century — is how it results in totalitarianism, colonialism, and genocide. As Zygmunt Bauman explained in *Modernity and the Holocaust* (1989), when "the modernist dream is embraced by an absolute power able to monopolize modern vehicles of rational action, and when that power attains freedom from effective social control, genocide follows".[6] This is why the so-called chaos brought about by the voiding of metanarratives through postmodernity did not aim to create a new order, as many today believe, but to avoid the external imposition of order. An open society, as Karl Popper explained while he was exiled in New Zealand and Europe was falling to authoritarian regimes, is one "in which individuals are confronted with personal decisions" as opposed to a "magical or tribal or collectivist society".[7] In the former, no one is in possession of the ultimate truth because it is acknowledged that people have different views, interests, and values. In the latter, truth is imposed by the bearers of power. Conway's insistence on alternative facts is a move in the ongoing return to order, a demonstration of the imposition of truth through power.

Trump, Peterson, and others who insist on these universal truths are creating a condition where the absence of emergency at the centre of this condition is the belief that there are no alternatives to the framed global order. This order imposes realism politically on other cultures and justifies its imposition intellectually by discrediting facts. Trump's hostility toward the facts of climate change, for example, is meant to create a condition without emergencies — where truth is imposed by authority, and nothing, not observations of the external world nor actions that would counteract the power of those authorities, is permitted to emerge from the overwhelming order. Difference, change, and cultural others must be avoided as disruptions of the safe situation that order claims to represent.

This order reveals itself every day as more authoritarian. Like Bernard Stiegler's "epoch of the absence of epoch",[8] the absence of emergency has become the greatest danger we face today, signaling the abandonment of the interpretative nature of existence in favor of the return to order and realism. But if, as Friedrich Hölderlin said, "where the danger is, also grows the saving power",[9] we must find ways to experience this danger, to reveal emergency from within its imposed absence.

Art often works better than scientific announcements and philosophical treatises as a way to reveal emergencies. This is not because of artists' ability to create beauty, but rather thanks to the intensity and depth of their works. Documentary photographs of the melting arctic, for example, can be truthful, but are rarely as powerful as works of art that address this emergency. "When a work of art truly takes hold of us", as Hans-Georg Gadamer said, "it is not an object that stands opposite us which we look at in the hope of seeing through it to an intended conceptual meaning. Just the reverse. The work is an *'Ereignis'*— an event that 'appropriates us' into itself. It shocks us, it overturns us, and sets up a world of its own, into which we are drawn".[10] Scientists and philosophers can also overturn our world, but their works preserve a distance that is constitutive of their findings and renders their effects less immediate. A work of art seeks to reduce this distance not only to draw our attention but also to involve us in an experience that the artist considers significant

Artists today are closer to the hidden emergencies than many scientists and philosophers, because art has been more resistant to the return to order. Science and systematic thought seek to "rescue us *from* emergencies" improving and preserving our order, but art at its best attempts to "rescue us into emergencies", creating event and shock. This rescue not only reveals absent emergencies — climate change, the unemployment crisis, and surveillance technologies, whatever has been concealed by the rhetoric of power; it also becomes an emergency, that is, an experience of danger. The artists who seek this experience are the ones whose work demands public intervention in global emergencies that are concealed in the idea of their absence. Art's role in stirring our existence is vital to opposing the absence of emergency sought by the advocates of the return to order. This is particularly evident in Pekka Niittyvirta and Timo Aho's installation *Lines (57° 59´N, 7° 16´W)* (2018–19) (Figure 1), Josh Kline's *Unemployment exhibition* (2016) (Figure 2), and Dries Depoorter's *Jaywalking Frames* (2018) (Figure 3).

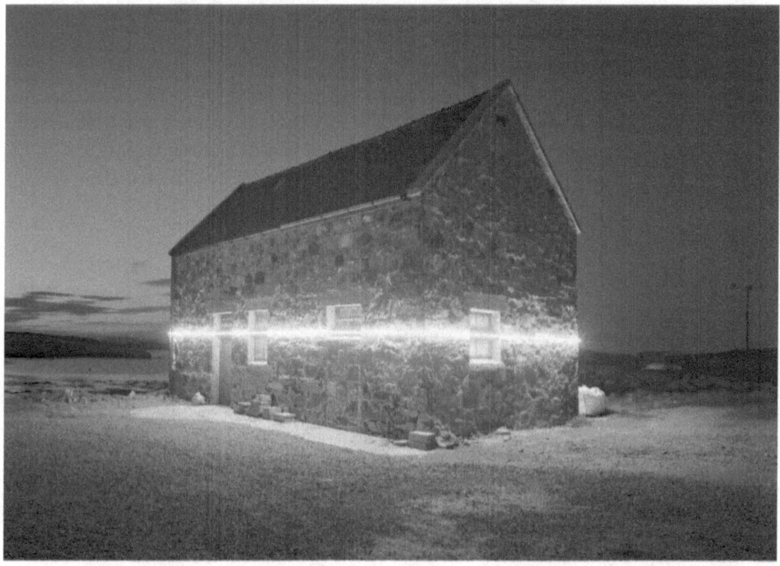

Figure 1. Pekka Niittyvirta and Timo Aho — *Lines (57° 59´N, 7° 16´W)*, installation view, Taigh Chearsabhagh Museum & Arts Centre, Lochmaddy, Scotland, 8 May 2018–31 August 2019.

In order to thrust us into the absent emergency of climate change, Niittyvirta and Aho installed a series of sensors on North Uist (in the Outer Hebrides of the Scottish islands) that activate synchronised beams of light that represent a scientific estimate of the level that the sea could rise to if the earth continues to warm. The public is not only invited to envision the emergency of the sea level increase, but also called to intervene and help mitigate the consequences of government indif-

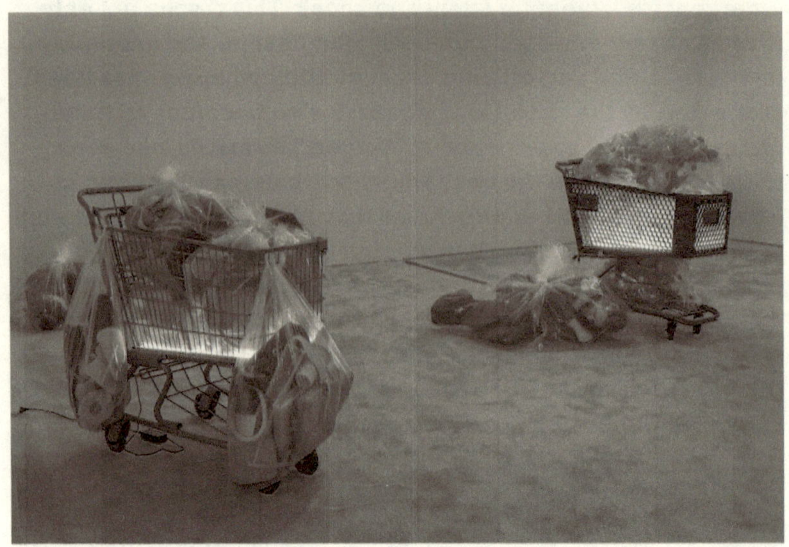

Figure 2. Josh Kline — *Unemployment,* 2016, installation, dimensions variable (Photo: Joerg Lohse). Courtesy the artist and 47 Canal, New York.

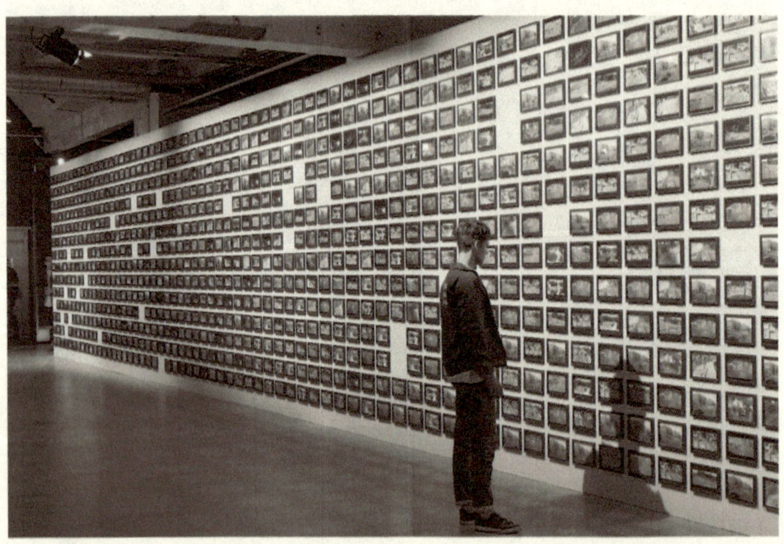

Figure 3. Dries Depoorter — *Jaywalking Frames,* installation view, MU Artspace, Eindhoven, The Netherlands, 20 July – 23 September 2018 (Photo: Boudewijn Bollmann).

ference. This indifference is also at the center of Kline's installation of life-size models of people in business attire tied up in garbage bags. Along with discarded office material (computer keyboards, family photos, and documents) these individuals represent people and businesses that disappeared in the wake of the mortgage crisis and ongoing processes of automation. Another emergency concealed at the heart of our contemporary societies is the multiplication of surveillance technologies. In order to thrust into the consequences of this emergency, Depoorter's work displays images of people walking through red lights that were captured through surveillance cameras and custom software. These technologies are meant not only to detect when traffic laws are broken, but also to develop "social credit systems" meant to assess each person's value in society.

These artists demand we intervene in environmental, social, and technological emergencies that we have not been able to confront because of the commonplace emergencies cited by the political return to order and realism. Their works, among others, are also a sign of the ongoing turn from "relational" to "emergency" aesthetic theories and artists' inevitable participation in global matters.[11] Although the art world, like scientific and political establishments, is also a system with hierarchies and frames, it has been affected by globalisation in a different way, one that through actual exchange lets works emerge in unusual locations and reveal different emergencies. The "globalization of the art world", as Arthur Danto once said, "means that art addresses us in our humanity, as men and women who seek in art for meanings that neither of art's peers — philosophy and religion — in what Hegel spoke of as the realm of Absolute Spirit, are able to provide".[12]

This is evident in the different experiences of art in art fairs and biennales: in static art fairs, the public contemplates works of art as valuable objects, whereas in the biennales the members of the audience all take responsibility for an experience that concerns everyone. In line with Gadamer's definition of a work of art, Caroline Jones believes it "is the emphasis on events and experiences, rather than objects, that constitute[s] the surprising legacy of biennial culture".[13] The fact that the latest trend in biennales, which have increased markedly in these past decades, is to offer these experiences in such remote places as Antarctica and the Californian desert is an indication that globalised art demands global interventions from its artists and its public.

In order to rescue the public *into* the greatest emergency — the imposed absence of emergency that is the result of an authoritarian return to order and realism — artists have begun to thrust us into this absence. These works demand from the public an intervention that is not only political, but also existential. Whether these artists will succeed in disrupting the "safe situation" politicians and thinkers are trying to impose will depend on the level of intervention from the public. Although these interventions are constantly undermined, the public must continue to question a return to order and realism, which have already proved tragic in the past.

Notes

1. Bari Weiss, "Meet the Renegades of the Intellectual Dark Web", *The New York Times,* 8 May 2018, https://www.nytimes.com/2018/05/08/opinion/intellectual-dark-web.html.

2. Jacob Hamburger, "The 'Intellectual Dark Web' *Is Nothing New", Los Angeles Review of Books,* 18 July 2018, https://lareviewofbooks.org/article/the-intellectual-dark-web-is-nothing-new.

3. Slavoj Žižek, *Sex and the Failed Absolute* (New York and London: Bloomsbury, 2019), 104–5.

4. Bruno Latour, *Down to Earth: Politics in the New Climate Regime,* trans. Catherine Porter (Cambridge: Polity, 2018), 23.

5. M. Gabriel, "Silicon Valley y las redes sociales son unos grandes criminales", interview with Ana Carbajosa in *El País,* 1 May 2019, https://elpais.com/cultura/2019/04/17/actualidad/1555516749_100561.html.

6. Zygmunt Bauman, *Modernity and the Holocaust* (London: Polity, 1989), 93–94.

7. Karl Popper, *The Open Society and Its Enemies* (London: Routledge & Sons, 1945), 172–173.

8. Bernard Stiegler, *The Age of Disruption,* trans. Daniel Ross (Cambridge: Polity, 2019), 5.

9. Friedrich Hölderlin, *Patmos* (Hamburg: Raamin-Presse, 1978 [1803]).

10. Hans-Georg Gadamer, *Gadamer in Conversation,* ed. R. E. Palmer (New Haven: Yale University Press, 2001), 71.

11. I refer to aesthetic theories such as Malcolm Miles's "eco-aesthetics", Jill Bennett's "practical aesthetics", and Veronica Tello's "counter-memorial aesthetics", where hidden emergencies such as the environment, terrorism, and refugees play central roles. On this turn, see the afterword in Santiago Zabala, *Why Only Art Can Save Us* (New York: Columbia University Press), 127–32.

12. Arthur Danto, *Unnatural Wonders: Essays from the Gap Between Art and Life* (New York: Columbia University Press, 2006), xvi.

13. Caroline A. Jones, *The Global Work of Art: World's Fairs, Biennials, and the Aesthetics of Experience* (Chicago: Chicago University Press, 2017), xiv–xv.

References

Bauman, Zygmunt. *Modernity and the Holocaust.* London: Polity, 1989.

Danto, Arthur. *Unnatural Wonders: Essays from the Gap Between Art and Life.* New York: Columbia University Press, 2006.

Gabriel, M. "Silicon Valley y las redes sociales son unos grandes criminales". Interview with Ana Carbajosa in *El País,* 1 May 2019. https://elpais.com/cultura/2019/04/17/actualidad/1555516749_100561.html.

Hans- Gadamer, *Georg. Gadamer in Conversation.* Edited by R. E. Palmer. New Haven: Yale University Press, 2001.

Hamburger, Jacob. "The 'Intellectual Dark Web' Is Nothing New". *Los Angeles Review of Books,* 18 July 2018. https://lareviewofbooks.org/article/the-intellectual-dark-web-is-nothing-new.

Hölderlin, Friedrich. *Patmos.* Hamburg: Raamin-Presse, 1978 [1803].

Jones, Caroline A. *The Global Work of Art: World's Fairs, Biennials, and the Aesthetics of Experience.* Chicago: Chicago University Press, 2017.

Latour, Bruno. *Down to Earth: Politics in the New Climate Regime.* Translated by Catherine Porter. Cambridge: Polity, 2018.

Popper, Karl. *The Open Society and Its Enemies.* London: Routledge & Sons, 1945.

Žižek, Slavoj. *Sex and the Failed Absolute.* New York and London: Bloomsbury, 2019.

Stiegler, Bernard. *The Age of Disruption.* Translated by Daniel Ross. Cambridge: Polity, 2019.

Weiss, Bari. "Meet the Renegades of the Intellectual Dark Web". *The New York Times,* 8 May 2018. https://www.nytimes.com/2018/05/08/opinion/intellectual-dark-web.html.

Zabala, Santiago. *Why Only Art Can Save Us.* New York: Columbia University Press, 2017.

Biographies

Bill Balaskas is an artist, theorist, and educator, whose research is located at the intersection of politics, digital media, and contemporary visual culture. He is an Associate Professor and Director of Research, Business and Innovation at the School of Art and Architecture of Kingston University, London. His work has been widely exhibited in the UK and internationally. He has received awards and grants from the European Investment Bank Institute; Comité International d'Histoire de l'Art (CIHA); Open Society Foundations; European Cultural Foundation; the Australian National University; and the Association for Art History (UK), amongst others. He has worked as an Editor of the *Leonardo Electronic Almanac* (MIT Press), and he is co-editor of *Institution as Praxis — New Curatorial Directions for Collaborative Research* (Sternberg Press, 2020); and *Architectures of Education* (e-flux Architecture, 2020). Originally trained as an economist, he holds a PhD in Critical Writing in Art and Design from the Royal College of Art, London.

Carolina Rito is Professor of Creative Practice Research, at the Research Centre for Arts, Memory and Communities (CAMC), Coventry University, UK, and lead of the Critical Practices research strand. She is a researcher and curator whose work explores "the curatorial" as an investigative practice, expanding practice-based research in the fields of curating, visual cultures, and cultural studies. Rito is Executive Board Member of the Midlands Higher Education & Culture Forum (MHECF); Research Fellow at the Institute of Contemporary History, Universidade NOVA Lisboa; Founding Editor of *The Contemporary Journal;* and Chair of the Collaborative Research Working Group for the MHECF. Rito is the co-editor of *Institution as Praxis—New Curatorial Directions for Collaborative Research* (Sternberg Press, 2020); Architectures of Education (e-flux Architecture, 2020); and editor of "On Translations" and "Critical Pedagogies" issues (*The Contemporary Journal,* 2018–2020). From 2017 to 2019, she was Head of Public Programmes and Research at Nottingham Contemporary. She holds a PhD in Curatorial/Knowledge from Goldsmiths, University of London.

Charlie Gere is Professor of Media Theory and History at Lancaster Institute for Contemporary Art, Lancaster University. He is the author of *Digital Culture* (2002–2008), *Art, Time and Technology* (2006); *Community Without Community in Digital Culture* (2012); *Unnatural Theology: Religion, Art and Media after the Death of God* (2019); and *I Hate the Lake District* (2020); and co-editor of *White Heat Cold Logic: British Computer Art 1960–1980* (2008), and *Art Practice in a Digital Culture* (2010).

Christine Ross is Professor and Distinguished James McGill Professor in Contemporary Art History at McGill University, Montreal. Her areas of research include contemporary media arts; vision and visuality; transformations of spectatorship in contemporary art; participatory media and art; artistic redefinitions of the public sphere; and reconfigurations of time and temporality in recent media art practices. She is currently working on a book project titled *Coexistence(s) in 21st Century Art* — a study on contemporary art's response to the migration crisis. Her recent publications include the co-edited volume, *The Participatory Condition in the Digital Age* (University of Minnesota Press, 2016), and *The Past is the Present; It's the Future too: The Temporal Turn in Contemporary Art* (Continuum, 2012). Between 2005 and 2018, she was the Principal Investigator of the FQRSC-funded MediaTopia team research projects. She was the Director of Media@McGill — a hub of interdisciplinary research, scholarship, and public outreach on issues in media, technology, and culture, from 2012 to 2017.

David M. Berry is Professor of Digital Humanities at University of Sussex and Visiting Fellow at *Forschungsverbund "Die Herausbildung normativer Ordnungen", Forschungs-kolleg Humanwissenschaften* (Institute for Advanced Studies), Goethe University Frankfurt am Main. His most recent books are *Digital Humanities: Knowledge and Critique in a Digital Age* (Polity, 2017) and *Critical Theory and the Digital* (Bloomsbury Academic, 2014). He has also recently co-edited *Postdigital Aesthetics: Art, Computation and Design* (Palgrave MacMillan, 2015).

Emily Rosamond has a research interest in how representations of self-worth are intertwined with financial infrastructures, imperatives, and understandings. She has written on a wide range of topics related to financialisation, subjectivity, and culture — and, most recently, Steve Bannon and the weaponisation of online reputation (*Theory, Culture & Society,* 2019); socially engaged art and the financialisation of social impact (*Finance and Society,* 2016); and online dating, addressivity, and assetised selves (*Journal of Aesthetics & Culture,* 2018). She is Lecturer in Visual Cultures at Goldsmiths, University of London, where she serves as Department Chair of Learning and Teaching. Her first monograph, *Reputation Warfare,* is forthcoming from Zone Books.

Forensic Architecture (FA) is a research agency, based at Goldsmiths, University of London. We undertake advanced spatial and media investigations into cases of human rights violations, with and on behalf of communities affected by political violence, human rights organisations, international prosecutors, environmental justice groups, and media organisations. We investigate state and corporate violence, human rights violations, and environmental destruction all over the world. Our work often involves open-source investigation, the construction of digital and physical models, 3D animations, virtual reality environments, and cartographic platforms. Within these environments we locate and analyse photographs, videos, audio files, and testimonies to reconstruct and analyse violent events. We also use our digital models as tools for interviewing survivors of violence, finding new ways to access and explore memories of trauma.

Gregory Sholette is an artist, writer, activist, and founding member of Political Art Documentation/Distribution (1989–1988); REPOhistory (1989–2000); and Gulf Labor Coalition (2010–ongoing). He recently edited a special double issue of *FIELD Journal of Socially Engaged Art* with over thirty global reports focusing on "Art, Anti-Globalism, and the Neo-Authoritarian Turn", and is co-editor with Chloë Bass of *Art as Social Action* (Skyhorse Publishers, 2018); as well as author of *Delirium and Resistance: Activist Art and the Crisis of Capitalism* (2017: Pluto Press); and *Dark Matter: Art and Politics in an Age of Enterprise Culture* (2010: Pluto Press). A graduate of the Whitney Independent Studies Program in Critical Theory, he holds an MFA from UC San Diego; BFA from The Cooper Union, and received his PhD from the University of Amsterdam, The Netherlands (2017). Sholette is affiliated faculty of the Art, Design and the Public Domain program of Harvard University's Graduate School of Design, and a Professor at Queens College CUNY, where he co-runs Social Practice Queens with Chloë Bass.

Mieke Bal has a commitment to interdisciplinary approaches to cultural artifacts and their potential effects. She focuses on gender, migratory culture, psychoanalysis, and the critique of capitalism. Her forty-some books include a trilogy on political art: *Endless Andness* (Bloomsbury Academic, 2013), *Thinking in Film* (Bloomsbury Academic, 2013), and *Of What One Cannot Speak* (University of Chicago Press, 2010). *Emma & Edvard Looking Sideways* (Yale University Press, 2017) demonstrates her integrated approach to academic, artistic, and curatorial work. She co-made documentaries on migratory culture, and "theoretical fictions". *A Long History of Madness* (2011) argues for a more humane treatment of psychosis, and was exhibited in a site-specific version, *Saying It,* at the Freud Museum London (2012). *Madame B* was combined with paintings by Edvard Munch in the Munch Museum in Oslo (2017). *Reasonable Doubt* explores the social aspects of thinking (2016). She exhibits a sixteen-channel video work *Don Quixote: tristes figuras.* Her latest film, *It's About Time! Reflections on Urgency* was produced in Poland (2020).

Nat Muller is an independent curator and writer with expertise in contemporary art from the Middle East. She has published widely on the topic and has curated numerous exhibitions and screenings for, among others, Eye Filmmuseum, and Stedelijk Museum, both in Amsterdam; Qalandiya International, Ramallah; Delfina Foundation, London; ifa-Galerie Berlin; The Mosaic Rooms, London; *International Film Festival Rotterdam; Kortfilmfestivalen/The Norwegian Short Film Festival;* and *International Short Film Festival Oberhausen.* In 2019, she curated the Danish Pavilion for the *58th Venice Biennale.* Her AHRC-funded PhD project at Birmingham City University researches science fiction in contemporary art from the Middle East.

Ferry Biedermann is a journalist and author who has worked extensively as a correspondent in the Middle East and Europe for, among others, the *Financial Times* and the Dutch daily newspaper, *de Volkskrant.* Since 2018, he has been writing on Brexit for Dutch daily *Trouw* and Birmingham City University's Centre for Brexit Studies. He is also editor-in-chief of a Dutch community quarterly magazine. His writings on journalism include his essay, "Being There: Journalists and Dead Bodies in Conflict", in *Bodies of Evidence* (Passagen Verlag, 2018). It draws on his experience reporting from Israel/Palestine, Iraq, Lebanon, and other areas of conflict. His work has appeared in, among others, *Foreign Policy, The Washington Post, The Daily Telegraph,* and Salon.com. He is a board member at the Dutch Association of Journalists, NUJ Netherlands, where he focuses on the position of freelancers.

Natalie Bookchin is an artist working at the intersection of media, politics, and internet culture. Her work has been exhibited extensively, including at MoMA PS1, Whitney Museum of American Art, and The Kitchen, all in New York; Centre Pompidou, Paris; and La Virreina Image Center, Barcelona. She has received numerous awards, including, among others, those from California Arts Council; Guggenheim Museum; The Rockefeller Foundation Bellagio Center; California Community Foundation; Jerome Foundation; Daniel Langlois Foundation; and MacArthur Foundation. She has received commissions from the Tate; Creative Time; LACMA | Los Angeles County Museum of Art; and Walker Art Center; among others. Bookchin is Professor at Rutgers University and lives and works in Brooklyn, NY.

Alexandra Juhasz is Distinguished Professor of Film at Brooklyn College, CUNY. She makes and studies committed media practices that contribute to political change and individual and community growth. Her current books are with Yvonne Welbon, *Sisters in the Life: 25 Years of African-American Lesbian Filmmaking* (Duke, 2018); with Jih-Fei Cheng and Nishant Shahani, *AIDS and the Distribution of Crises* (Duke 2020); with Nishant Shah, *Really Fake!* (University of MN and Melos Presses, 2020); and *My Phone Lies to Me: Fake News Poetry Workshops as Radical Digital Media Literacy* (currently seeking a press). She currently works on, and about, feminist Internet culture, including fake news (http://scalar.me/100hardtruths) and *Fake News Poetry Workshops* (http://fakenews-poetry.org). With Anne Balsamo, she was founding co-facilitator of the network, FemTechNet (https://femtechnet.org). Her most recent work is the podcast: We Need Gentle Truths for Now: (https://shows.acast.com/we-need-gentle-truths-for-now).

Ramon Bloomberg is a writer and film-maker based in London. His research examines the impact of technical development and new technical milieux on the political status of the liberal subject. Bloomberg writes fiction under the pen name, Jack Lively.

Santiago Zabala is ICREA Research Professor of Philosophy at Pompeu Fabra University, Barcelona, and author of many books, including *Why Only Art Can Save Us: Aesthetics and the Absence of Emergency* (Columbia University Press, 2017). He has written for *The Guardian, New York Times,* and *Al-Jazeera.* His most recent book is *Being at Large: Freedom in the Age of Alternative Facts* (McGill-Queen's University Press, 2020).

Steven Henry Madoff is the founding chair of the Masters in Curatorial Practice program at the School of Visual Arts in New York. Previously, he served as senior critic at Yale University's School of Art. He lectures internationally on such subjects as the history of interdisciplinary art, contemporary art, and art pedagogy. He has curated exhibitions internationally over the last thirty years in the United States, Europe, and the Middle East. His art criticism, journalism, and theoretical texts have been translated into many languages. Madoff's most recent book is *What About Activism?* (editor) from Sternberg Press, 2019. He holds his PhD in Modern Thought and Literature from Stanford University.

Terry Smith is Andrew W. Mellon Professor of Contemporary Art History and Theory in the Department of the History of Art and Architecture at the University of Pittsburgh, and Professor in the Division of Philosophy, Art and Critical Thought at the European Graduate School. He is also Lecturer at Large, Curatorial Program, School of Visual Arts, New York. Smith is author of *Making the Modern: Industry, Art and Design in America* (University of Chicago Press, 1993); *Transformations in Australian Art (*Craftsman House, 2002); *The Architecture of Aftermath* (University of Chicago Press, 2006); *What is Contemporary Art?* (University of Chicago Press, 2009); *Contemporary Art: World Currents* (Laurence King and Pearson Prentice Hall, 2011); *Thinking Contemporary Curating* (Independent Curators International, 2012); *Talking Contemporary Curating* (Independent Curators International, 2015); *The Contemporary Composition* (Sternberg Press, 2016); *One and Five Ideas: On Conceptual Art and Conceptualism* (Duke University Press, 2107); and *Art to Come: Histories of Contemporary Art* (Duke University Press, 2019).

UBERMORGEN (AT/CH/US) is an artist duo founded in Vienna in 1995 by lizvlx and Hans Bernhard. Part of the Net.Art avant-garde of the 1990s and 2000s digital actionism and concept art, UBERMORGEN celebrate a radical-subversive approach to data and matter. UBERMORGEN owns 175 websites/domains, and they have been featured in more than 3000 news reports and reviews. CNN calls them "Maverick Austrian Business People", and *New York Times* calls them "simply brilliant". UBERMORGEN has exhibited at Centre Pompidou, Paris; *Liverpool Biennial,* UK; Serpentine Galleries, London; MoMA PS1, Whitney Museum of American Art, and New Museum, all in New York; *Biennale of Sydney,* AU; Barcelona Museum of Contemporary Art (MACBA); San Francisco Museum of Modern Art (SFMOMA); InterCommunication Centre (ICC), Tokyo; and *Gwangju Biennale,* South Korea. Their main influences include: Rammstein, Samantha Fox, XXXTentacion, and Pixi-Bücher, Olanzapine, and LSD, Kentucky Fried Chicken's Coconut Shrimp Deluxe, and Viennese Actionism. UBERMORGEN talk at international conferences, museums, and symposia, and they hold the Professorship in Networks at the Academy of Media Arts Cologne (KHM).

www.ingramcontent.com/pod-product-compliance
Lightning Source LLC
LaVergne TN
LVHW091125080826
845145LV00008B/2046